TRANSLATIONS

OF

MATHEMATICAL

MONOGRAPHS

Volume 25

MAHLER'S PROBLEM IN METRIC NUMBER THEORY

by

V. G. Sprindžuk

AMERICAN MATHEMATICAL SOCIETY

Providence, Rhode Island 02904

1969

ПРОБЛЕМА МАЛЕРА
В МЕТРИЧЕСКОЙ ТЕОРИИ ЧИСЕЛ

В. Г. СПРИНДЖУК

Издательство „Наука и техника“
Минск 1967

Translated from the Russian by
B. Volkmann

Library of Congress Card Number 73–86327
Standard Book Number 821–81575–X

Printed in the United States of America

PREFACE

This book deals with the solution of a group of questions related both to the general theory of transcendental numbers and to the metrical theory of diophantine (and also algebraic) approximations. The fundamental problem in this field has been known in the literature since 1932 as Mahler's conjecture, since it arose in connection with the classification of numbers which he introduced (cf. [41],[42]).

After Mahler's basic papers, a number of mathematicians (Koksma, LeVeque, Kubilius, Kasch, Volkmann and others) achieved significant advances in their efforts to prove Mahler's conjecture.

The main result of this book is a proof of Mahler's conjecture and some analogous theorems (on p-adic numbers and on power series over finite fields). A complete account of the proof of the conjecture is to be found in the author's paper [68]; a short exposition of the method of proof is given in [65], [66], and [67]. Auxiliary and preliminary results have appeared in [55], [56], and [59].

In Part I we consider the "classical" case of Mahler's conjecture, dealing with real and complex numbers. Part II is concerned with locally compact fields with non-archimedean valuation (fields of p-adic numbers and of formal power series over finite fields).

The basic idea underlying the method of approach is the same in both parts, and the final results are analogous. The proofs given are "minimal" in the sense that they involve only such concepts and facts which are indispensible for the essence of the subject.

The part on "Supplementary results and remarks" deals with problems which are not directly related to the main theme but have, nevertheless, close connections with Mahler's conjecture.

The author hopes that the exposition in Part I will be comprehensible to anyone who knows the elements of measure theory and possesses sufficient preseverance in overcoming purely logical difficulties. Beyond this, Part II requires a general familiarity with the structure of fields with non-archimedean valuation. All the necessary information is given in the text with references to the sources.

In the Appendix detailed proofs of some new theorems in the metrical theory of diophantine approximation are given.

I wish to express my heart-felt thanks to J. P. Kubilius for having introduced me to this interesting branch of number theory and to Ju. V. Linnik for his support of my first efforts. I also wish to thank B. F. Skubenko who devoted much energy to a careful reading and checking of the proof of Mahler's conjecture. I take pleasure also in expressing my thanks to Professor K. Mahler for his cordial congratulations in connection with the verification of his time-honored conjecture. Finally, I want to thank all the members of the Institute of Mathematics of the Academy of Sciences of the Belorussian Soviet Socialist Republic who have helped to prepare the publication of this book.

V. Sprindžuk

TABLE OF CONTENTS

INTRODUCTION

§1. BASIC CONCEPTS

Let ω be a real or complex number. We will consider the approximation of the number zero by the values at the point ω of polynomials

$$P(x) = a_0 + a_1 x + \ldots + a_n x^n$$

with integer coefficients, with fixed height

$$h(P) = \max(|a_0|, |a_1|, \ldots, |a_n|)$$

and given degree n $(n = 1, 2, \cdots)$.

We define $w_n(\omega, H) = \min |P(\omega)|$, where the minimum is taken with respect to all polynomials P with integer coefficients, of degree less than or equal to n, with height less than or equal to H and with the property $P(\omega) \neq 0$. Furthermore we introduce the parameters

$$w_n(\omega) = \overline{\lim_{H\to\infty}} \frac{\ln \dfrac{1}{w_n(\omega, H)}}{\ln H},$$

and

$$w(\omega) = \overline{\lim_{n\to\infty}} \frac{1}{n} w_n(\omega).$$

Evidently, if ω is transcendental, $w_n(\omega)$ is the supremum of the set of all $w > 0$ for which the inequality

$$|P(\omega)| < h_P^{-w}, \quad h_P = h(P), \tag{1}$$

is solved, as h_P tends to infinity, by infinitely many such polynomials P of degree n.

It follows immediately that always $0 \leq w_n(\omega) \leq w_{n+1}(\omega) \leq \infty$ and $0 \leq w(\omega) \leq \infty$. If $w(\omega) = \infty$ and if there exists an index $\mu = \mu(\omega)$ with $w_\mu(\omega) = \infty$, then let μ be the smallest index for which this is true. Otherwise we define $\mu(\omega) = \infty$.

In 1932 Mahler [41] introduced a classification in which the set of complex numbers is divided into the following classes: The number ω is called

$$\begin{array}{lll} \text{an } A\text{-number if} & w(\omega) = 0, & \mu(\omega) = \infty, \\ \text{an } S\text{-number if} & 0 < w(\omega) < \infty, & \mu(\omega) = \infty, \\ \text{a } T\text{-number if} & w(\omega) = \infty, & \mu(\omega) = \infty, \\ \text{a } U\text{-number if} & w(\omega) = \infty, & \mu(\omega) < \infty. \end{array}$$

It is easy to verify that this list of classes contains all logically possible combinations of values of $w(\omega)$ and $\mu(\omega)$. Furthermore, it can be shown that the class of A-numbers consists of the algebraic numbers; consequently, the transcendental numbers form the classes of S-, T-, and U-numbers.

It is easily shown by means of Dirichlet's "pigeonhole principle" that, for any transcendental number ω, there exists an infinite sequence of polynomials P of degree less than or equal to n, with integer coefficients, which satisfy the inequality

$$|P(\omega)| < c(n, \omega) h_P^{-w},$$

where $w = n$ if ω is real, and $w = (n-1)/2$ if ω is a nonreal complex number; furthermore $h_P = h(P)$ may be assumed to tend to infinity. Thus,

$$w_n(\omega) \geqslant \begin{cases} n & \text{if } \omega \text{ is real} \\ \dfrac{n-1}{2} & \text{if } \omega \text{ is complex.} \end{cases} \tag{2}$$

Hence $w(\omega) \geq 1$ if ω is a real transcendental number, and $w(\omega) \geq ½$ for any complex transcendental number.

It has been shown (see [41]) that the number e, the basis of the natural logarithms, is an S-number satisfying $w_n(e) = 1$ $(n = 1, 2, \cdots)$. Furthermore, it can be shown by construction (cf. [37]) that there exist U-numbers for every natural index μ. In particular, Liouville's numbers turn out to be U-numbers of index one. It is still unknown at present whether T-numbers exist.

In the sequel we shall always assume ω to be transcendental. It is convenient to introduce the following notation:

$$\frac{1}{n} w_n(\omega) = \begin{cases} \Theta_n(\omega) \text{ if } \omega \text{ is real,} \\ \eta_n(\omega) \text{ otherwise}. \end{cases}$$

$$\Theta(\omega) = \sup_{(n)} \Theta_n(\omega) \quad (n = 1, 2, \ldots),$$

$$\eta(\omega) = \sup_{(n)} \eta_n(\omega) \quad (n = 2, 3, \ldots).$$

Thus the inequalities (2) may be written in the following form:

$$\Theta_n(\omega) \geqslant 1 \qquad (n = 1, 2, \ldots),$$

$$\eta_n(\omega) \geqslant \frac{1}{2} - \frac{1}{2n} \qquad (n = 2, 3, \ldots). \tag{3}$$

Mahler's classification is closely related to the classification of transcendental numbers due to Koksma [30]: Let $w_n^*(\omega, H) = \min|\omega - \kappa|$, where κ runs through all algebraic numbers whose degree and height do not exceed n and H, respectively. We define the height of an algebraic number κ as the height of its minimal polynomial, i.e. the primitive polynomial of lowest degree, with integer coefficients, which has κ as a zero. Furthermore we introduce

$$w_n^*(\omega) = \overline{\lim_{H\to\infty}} \frac{\ln \frac{1}{H w_n^*(\omega, H)}}{\ln H}$$

and

$$w^*(\omega) = \overline{\lim_{n\to\infty}} \frac{1}{n} w_n^*(\omega).$$

Evidently, $w_n^*(\omega)$ is the supremum of the set of all numbers $w > 0$ for which there exist infinitely many algebraic numbers κ of degree less than or equal to n and height $h_\kappa = h(\kappa)$ satisfying the inequality

$$|\omega - \kappa| < h_\kappa^{-1-w} \qquad (h_\kappa \to \infty). \tag{4}$$

Obviously, $0 \le w_n^*(\omega) \le w_{n+1}^*(\omega) \le \infty$ and $0 \le w^*(\omega) \le \infty$. We define μ^* as the smallest index with $w_{\mu^*}(\omega) = \infty$ if such an index exists; otherwise we let $\mu^* = \infty$.

Koksma called a number ω

an S^*-number if $w^* < \infty$, $\mu^* = \infty$,

a T^*-number if $w^* = \infty$, $\mu^* = \infty$,

a U^*-number if $w^* = \infty$, $\mu^* < \infty$.

It can be shown (see [30], [54]) that Mahler's classes S, T, U coincide with the corresponding classes of Koksma.

We let

$$\frac{1}{n} w_n^*(\omega) = \begin{cases} \Theta_n^*(\omega) & \text{if } \omega \text{ is real,} \\ \eta_n^*(\omega) & \text{otherwise.} \end{cases}$$

Wirsing [80] (see also [56]) found the following relations (which he expressed in terms of the quantities $w_n(\omega)$ and $w_n^*(\omega)$):

$$\Theta_n^*(\omega) \geqslant \frac{1}{2}\Theta_n(\omega) + \frac{1}{2n},$$

$$\Theta_n^*(\omega) \geqslant \frac{\Theta_n(\omega)}{n\left(\Theta_n(\omega) - 1 + \frac{1}{n}\right)} \qquad (n = 1, 2, \ldots), \tag{5}$$

$$\eta_n^*(\omega) \geqslant \frac{1}{2}\eta_n(\omega),$$

$$\eta_n^*(\omega) \geqslant \frac{\eta_n(\omega)}{n\left(2\eta_n(\omega) - 1 + \frac{2}{n}\right)} \qquad (n = 2, 3, \ldots). \tag{6}$$

In connection with these inequalities Wirsing expressed the first half of the conjecture that, for any transcendental number ω,

$$\Theta_n^*(\omega) \geqslant 1 \qquad \left(\eta_n^*(\omega) \geqslant \frac{1}{2} - \frac{1}{2n}\right) \qquad (n = 2, 3, \ldots) \tag{7}$$

depending on whether ω is real or not. By virtue of (3), Wirsing's inequalities (5) and (6) imply the relations

$$\Theta_n^*(\omega) \geqslant \frac{1}{2} + \frac{1}{2n}, \quad \eta_n^*(\omega) \geqslant \frac{1}{4} - \frac{1}{4n} \qquad (n = 2, 3, \ldots).$$

§2. HISTORICAL SURVEY

In 1932 it was shown by Mahler [42] that almost all real and almost all complex numbers (in the sense of linear and planar Lebesgue measure, respectively) are S-numbers. Furthermore, Mahler proved that there exists a constant $\gamma > 0$ such that for almost all numbers ω and for all n, every polynomial P of degree less than or equal to n, with integer coefficients and with height $h_P > h_0(\omega, n, \gamma)$, satisfies the inequality

$$|P(\omega)| > h_P^{-n\gamma}, \tag{8}$$

According to Mahler this is true with

$$\gamma = 4 + \varepsilon, \text{ and } \gamma = \frac{7}{2} + \varepsilon \tag{9}$$

in the real and complex cases, respectively, where $\epsilon > 0$ is an arbitrary constant. In the same paper Mahler stated the conjecture that in the assertion (8) the values given of (9) could be replaced by

$$\gamma = 1 + \varepsilon \quad \text{and} \quad \gamma = \frac{1}{2} + \varepsilon \tag{10}$$

respectively.

Thus, Mahler's result (9) on the measure of the set of S-numbers is equivalent to the inequalities

$$\Theta(\omega) \leq 4, \quad \eta(\omega) \leq \frac{7}{2}, \tag{11}$$

for almost all real and for almost all complex numbers, respectively. In the same sense it follows from (3) and (10) that Mahler's conjecture may be expressed by the two equations

$$\Theta(\omega) = 1 \text{ and } \eta(\omega) = \frac{1}{2}. \tag{12}$$

Subsequently, Mahler's inequalities (11) were improved. At first Koksma [30] obtained the inequalities

$$\Theta(\omega) \leq 3, \quad \eta(\omega) \leq \frac{5}{2},$$

for almost all real and almost all complex numbers. Later LeVeque [36], using a lemma of N. I. Fel'dman [13], proved that $\Theta(\omega) \leq 2$ and $\eta(\omega) \leq 3/2$ in the same sense. In still later years Kasch and Volkmann (cf. [24], [25], [74]) showed that, for almost all numbers ω,

$$\Theta_n(\omega) \leq 2 - \frac{2}{n} \quad \text{and} \quad \eta_n(\omega) \leq 1 - \frac{1}{n} \qquad (n = 2, 3, \ldots).$$

W. Schmidt [52], refining the methods of Kasch and Volkmann, obtained the inequalities

$$\Theta_n(\omega) \leq 2 - \frac{7}{3n} \qquad (n = 3, 4, \ldots)$$

for almost all numbers ω. Finally, Volkmann [76] has shown recently that

$$\Theta_n(\omega) \leq \frac{3}{2}, \quad \eta_n(\omega) \leq \frac{3}{4} - \frac{1}{2n} \qquad (n = 2, 3, \ldots),$$

for almost all numbers ω, and in a later paper [77] he sharpened this to

$$\Theta_n(\omega) \leq \frac{4}{3}, \quad \eta_n(\omega) \leq \frac{2}{3} - \frac{1}{2n} \qquad (n = 2, 3, \ldots).$$

This result was also obtained independently by the author [59]. More precisely, he proved the inequalities

$$\Theta_n(\omega) \leqslant \frac{5}{4} - \frac{3}{8n} \qquad (n = 2, 3, \ldots, 7),$$

$$\Theta_n(\omega) \leqslant \frac{4}{3} - \frac{1}{n} \qquad (n = 8, 9, \ldots),$$

$$\eta_n(\omega) \leqslant \frac{5}{8} - \frac{11}{16n} \qquad (n = 2, 3, \ldots, 7),$$

$$\eta_n(\omega) \leqslant \frac{2}{3} - \frac{1}{n} \qquad (n = 8, 9, \ldots)$$

for almost all numbers ω.

Together with the general results mentioned above, the exact values of the parameters $\Theta_n(\omega)$ and $\eta_n(\omega)$ for almost all ω were determined in the cases $n = 1, 2, 3$.

The equation $\Theta_1(\omega) = 1$ for almost all real numbers ω follows immediately from results of A. Ja. Hinčin[1)] [[21]] and [[19]]). The equation $\Theta_2(\omega) = 1$ for almost all real ω and also a sharper result were shown by J. P. Kubilius ([[31]] and [[32]]) by applying Vinogradov's method of trigonometric sums (cf. [[5]], [[23]], [[73]]). Finally, Volkmann [[72]] obtained the equality $\Theta_3(\omega) = 1$ for almost all real ω, basing his proof on results of Davenport ([[8]], [[9]]) on binary cubic forms.

In the case of complex numbers Kasch [[23]] proved that $\eta_2(\omega) = 1/4$ almost everywhere, and Volkmann [[73]] showed that $\eta_3(\omega) = 1/3$ for almost all ω (see also [[56]]).

It is interesting to note that Kasch [[23]] stated the following conjecture:

$$\eta_n(\omega) = \frac{1}{2} - \frac{1}{2n} \qquad (n = 2, 3, \ldots) \tag{13}$$

for almost all ω. In the cases $n = 2$ and $n = 3$ this conjecture is verified by the theorems of Kasch and Volkmann just mentioned.

The problem has been carried over to the case of p-adic numbers (cf. Mahler [[44]], Turkstra [[69]]) where the results were, by and large, analogous to those indicated above (cf. Turkstra [[69]], Lock [[38]], Kasch-Volkmann [[26]], Sprindžuk [[55]], [[64]]). Furthermore, an analogous problem has been studied for formal power

---

1) *Translator's note.* It also follows from older results.

series (Sprindžuk [59], [55]).

Let Q_p be the field of p-adic numbers, $\omega \in Q_p$, and let $|\omega|_p$ denote the p-adic value of ω in Q_p. For given positive integers n and H we define $w_n(\omega, H) = \min |F(\omega)|_p$, where the minimum is taken with respect to all polynomials F of degree less than or equal to n, with integer coefficients, whose height does not exceed H and for which $F(\omega) \neq 0$. Mahler considered the quantities

$$w_n(\omega) = \overline{\lim_{H\to\infty}} \frac{\ln \dfrac{1}{w_n(\omega, H)}}{\ln H},$$

and

$$w(\omega) = \overline{\lim_{n\to\infty}} \frac{1}{n} w_n(\omega),$$

and used them to introduce a classification of the numbers from Q_p similar to his classification in the real and in the complex case.

Lock [38] proved that almost all (in the sense of Turkstra's measure; cf. [69]) numbers from Q_p belong to Mahler's class of S-numbers. In precise terms, Lock's result amounts to the inequalities

$$n + 1 \leqslant w_n(\omega) \leqslant 3n + 1 \qquad (n = 1, 2, \ldots), \tag{13'}$$

Later Kasch and Volkmann [26] proved the inequalities

$$w_n(\omega) \leqslant 2n - \frac{1}{2} \qquad (n = 3, 4, \ldots)$$

and the equations $w_1(\omega) = 2$, $w_2(\omega) = 3$ for almost all $\omega \in Q_p$. Additional results were obtained by the author (see [55], [64]):

(1) There exist numbers w_n $(n = 1, 2, \cdots)$ such that

$$w_n(\omega) = w_n \qquad (n = 1, 2, \ldots)$$

for almost all $\omega \in Q_p$.

(2) These numbers satisfy the inequalities

$$w_n \leqslant \frac{5}{4}(n + 1) \qquad (n = 3, 4, 5, 6, 7),$$

$$w_n \leqslant \frac{4}{3} n \qquad (n = 8, 9, \ldots).$$

(3) The equation $w_3 = 4$ holds.

An attempt by Kasch and Volkmann [27] to prove statement (3) was fallacious (see R. Ž. Mat. 1962, Review 8A104). In analogy with Mahler's conjecture on real S-numbers one might expect that, for almost all $\omega \in Q_p$,

$$w_n(\omega) = n + 1 \quad (n = 1, 2, \ldots).$$

Finally, let K be a finite field, x a transcendental element over K; let $R = K[x]$ be the ring of polynomials in x over K, and $K\langle x\rangle$ the field of formal power series of x^{-1} over K, i.e. the field of all series of the form

$$\omega = \sum_{s=l}^{\infty} \alpha_s x^{-s}, \ \alpha_s \in K \quad (s = l, l+1, \ldots).$$

We introduce in $K\langle x\rangle$ a non-archimedean norm $|\omega|$ by defining

$$|\omega| = \begin{cases} 0, & \text{if } \alpha_s = 0 \text{ for all } s, \\ q^{-l}, & \text{if } l \text{ is the smallest index with } \alpha_l \neq 0, \end{cases}$$

where q denotes the number of elements of K. Then we define $w_n(\omega, H) = \min |P(\omega)|$, the minimum being taken with respect to all polynomials P over the ring R with $P(\omega) \neq 0$ whose degree and height do not exceed n and H, respectively. Here the height $h(P)$ of a polynomial

$$P(z) = a_0 + a_1 z + \ldots + a_n z^n, \ a_i \in R,$$

is defined in a natural way as

$$h(P) = \max(|a_0|, |a_1|, \ldots, |a_n|).$$

Finally, we let

$$w_n(\omega) = \overline{\lim_{H\to\infty}} \frac{\ln \dfrac{1}{w_n(\omega, H)}}{\ln H}.$$

The author [55], [60] proved that almost all (in the sense of the Haar measure in the field $K\langle x\rangle$) elements $\omega \in K\langle x\rangle$ satisfy the inequalities

$$n \leqslant w_n(\omega) \leqslant \frac{4}{3} n \quad (n = 1, 2, \ldots)$$

Again one might conjecture that, in analogy with the case of real numbers,

$$w_n(\omega) = n \quad (n = 1, 2, \ldots).$$

for almost all $\omega \in K\langle x\rangle$.

§3. GENERAL OUTLINE OF THE PROOF

This book is concerned with a proof of Mahler's conjecture, i.e. of the assertion that almost all numbers ω satisfy the equations (12). We shall also prove the conjectures (13) of Kasch and (7) of Wirsing. Analogous results are obtained for p-adic numbers and for power series over finite rings.

In the first part we shall deal with the cases of real and complex numbers. In particular, the latter case will serve as our starting point and will be discussed in greater detail than the case of real numbers. In the second part of the book we develop the basic facts which are necessary for the study of analogs to Mahler's hypothesis in the case of locally compact non-archimedean normed rings and which are suited for the proof of the natural analogs in the cases of p-adic numbers and of power series.

In both parts of the book the exposition is based on analogous considerations. Therefore, we found it appropriate to explain here the crux of our method of proof, using the cases of complex and real numbers as a basis.

The proof of Mahler's conjecture breaks down in a natural way into two steps:

1) the existence of numbers w_n such that

$$w_n(\omega) = w_n \quad (n = 1, 2, \ldots)$$

for almost all ω;

2) the equations

$$w_n = (n-1)/2 \quad (n = 2, 3, \cdots) \text{ and } w_n = n \quad (n = 1, 2, \cdots)$$

in the complex and real cases, respectively.

The first result will be established easily by means of Lemma 1 (Part I, Chapter 1, §4). The main fact here is the invariance of the function $w_n(\omega)$ under fractional linear transformations with integer coefficients and the property of measurable sets described by Lemma 11.

The second result is more significant and represents the principal assertion of the conjecture. As it turns out, a proof by induction on n is possible. Formally, this induction consists in establishing the inequalities

$$w_n \leqslant \frac{1}{2} + w_{n-1} \quad (n = 3, 4, \ldots) \tag{14}$$

in the complex case, and

$$w_n \leqslant 1 + w_{n-1} \quad (n = 2, 3, \ldots) \tag{15}$$

in the case of real numbers. Once this has been accomplished, the assertion of the theorem follows from (3) and a computation of the known quantities w_2.

The proof of the relations (14) and (15) is preceded by certain preparatory work. Here the most important part is the reduction of (1) to a situation where only irreducible primitive polynomials satisfying condition (21), with integer coefficients, have to be admitted (Part I, Chapter 1, §§5 and 6), i.e. the reduction to polynomials from the class $\mathbf{P}_n$ in the terminology of §6.

Let $w_0 = w_{n-1} + \delta$, where $\delta > 0$ is an arbitrary, but henceforth fixed, real number. For any polynomial $P \in \mathbf{P}_n$ we introduce a system of simply connected domains $\sigma_i(P)$ (in the real case, of intervals) in which $|P(\omega)| < h_p^{-w_0}$.

The domains $\sigma_i(P)$ are subdivided into two classes, called essential and inessential. In the former case $\sigma_i(P)$ contains a set of measure greater than $\frac{1}{2}$ meas $\sigma_i(P)$ such that none of its points belongs to any other domain $\sigma_j(Q)$ for any polynomial $Q \in \mathbf{P}_n$ with the same height as P. If a domain $\sigma_i(P)$ does not have this property then it is called inessential. The vital fact here is that, for any $\delta > 0$, the set of all points contained in infinitely many inessential domains has measure zero. This follows from Lemmas 10 and 9, respectively, by means of results of Part I, Chapter 2, §1.

What remains to be done after this step is to introduce the supremum of those $w > 0$ for which the inequality (1) is, for almost all ω, satisfied by at most finitely many polynomials $P \in \mathbf{P}_n$. Here we may now assume that ω belongs to some essential domain (or interval). In order to be able to use this property we make the transition from (1) to the system (77) (Part I, Chapter 1, §7). The investigation of the system (77) proceeds by dividing the algebraic numbers of degree n into classes $K(\mathrm{r})$ and by subsequently applying the necessary arguments to each class separately. It turns out to be appropriate to distinguish between two types of classes $K(\mathrm{r})$: classes of the first and classes of the second kind (Part I, Chapter 2, §4). The study of the system (119) for classes of the first kind is based on the properties of essential domains $\sigma_i(P)$ corresponding to the roots $\kappa \in K(\mathrm{r})$ of a polynomial P. In this case a relation of the form (14) or (15) is obtained.

The classes of the first kind contain the vast majority of algebraic numbers κ of degree n, whereas the classes of the second kind are comprised of only a small portion of these numbers. The latter ones are treated by a modified version of a method due to Volkmann [76], [77] and the author [55], [59] which is independent of the concept of essential and inessential domains.

The exposition in the second part of the book follows the same outline.

Part I

REAL AND COMPLEX NUMBERS

CHAPTER 1

AUXILIARY CONSIDERATIONS

§1. NOTATION

In the sequel $P(x)$ denotes a polynomial

$$P(x) \doteq a_0 + a_1x + \ldots + a_nx^n, \quad n \geqslant 2 \tag{16}$$

of the formal degree n. The coefficients $a_0, a_1, \cdots, a_n$ will generally be supposed to be integers or, sometimes, real numbers. Both cases will be indicated in advance.

Instead of the polynomial $P(x)$ we will sometimes consider the polynomial

$$P_{(l)}(x) = P(x+l) \tag{17}$$

where l is an integer, the polynomial

$$\overline{P}(x) = x^n a_0 + x^{n-1}a_1 + \ldots + a_n = x^n P\left(\frac{1}{x}\right), \tag{18}$$

and also the polynomial $\overline{P}_{(l)}(x)$ which is obtained from $P(x)$ by applying the transformation (17) followed by the transformation (18).

The zeros of the polynomial (16) will be denoted by $\kappa_1, \kappa_2, \cdots, \kappa_n$ (here we assume $a_n \neq 0$). The discriminant of the polynomial (16) will be written as $D(P)$. We shall often use the product representation of the discriminant, i.e. the equation

$$D(P) = a_n^{2n-2} \prod_{1 \leqslant i < j \leqslant n} (\varkappa_i - \varkappa_j)^2. \tag{19}$$

The resultant of two polynomials $P_1(x)$, $P_2(x)$ will be denoted by $R(P_1, P_2)$.

With each zero κ_j of the polynomial (16) we associate the set $S(\kappa_j)$ of complex numbers ω which satisfy the condition

$$|\omega - \kappa_j| = \min_{(i)} |\omega - \kappa_i| \quad (i = 1, 2, \dots, n), \tag{20}$$

i.e. the set of all complex numbers whose distance from κ_j is not greater than their distance from the other zeros $\kappa_1, \kappa_2, \cdots, \kappa_n$. Clearly, $S(\kappa_j)$ is a closed point set whose boundary consists of line segments.

By and large we shall deal with polynomials (16) without multiple zeros, subject to the condition

$$\max(|a_0|, |a_1|, \dots, |a_{n-1}|) \leqslant a_n. \tag{21}$$

Obviously, $h(P) = a_n$ for any such polynomial. The set of these polynomials will be denoted by P_n. Let $\mathrm{P}_n(h)$ be the subset of polynomials $P \in \mathrm{P}_n$ for which $h(P) = h$. Hence $\mathrm{P}_n(h)$ consists of the polynomials (16) without multiple zeros for which

$$\max(|a_0|, |a_1|, \dots, |a_{n-1}|) \leqslant a_n = h.$$

The range of definition of the coefficients will be specified later.

We will use the notation $c(n)$ for positive functions of n, and we will use the formal rules

$$c(n) + c(n) = c(n), \quad c(n)\, c(n) = c(n),$$

with the obvious meaning that the sum and the product of positive functions of n are again positive functions of n.

Sometimes we will use the symbol $\ll$, as introduced by I. M. Vinogradov, in the following sense: If X and Y are positive parameters then $X \ll Y$ means that there exists a constant c, independent of X and Y, such that $X < cY$.

Finally, if $\{A_i\}_{i=1,2\cdots}$ is a denumerable system of mutually disjoint sets A_i, we shall denote the union $A = \bigcup_{i=1}^{\infty} A_i$ by the summation symbol

$$A = \sum_{i=1}^{\infty} A_i. \tag{22}$$

Conversely, the notation (22) means that $A = \bigcup_{i=1}^{\infty} A_i$ and that the sets A_i are mutually disjoint.

All further definitions and abbreviations will be introduced later on.

§2. LEMMAS ON POLYNOMIALS

In this section the coefficients of the polynomials $P(x)$ will be subjected only to the condition that they be real numbers, unless otherwise specified.

Lemma 1. *If there exists a number* c, $0 < c \leq 1$, *such that the polynomial* $P(x)$ *satisfies the condition* $|a_n| > ch$, $h = h(P)$, *then*

$$\max_{i=1,2,\dots n} |\varkappa_i| \leqslant \frac{n}{c}. \tag{23}$$

Proof. Let κ be any zero of the polynomial $P(x)$. If $|\kappa| > 1$, then the equation

$$a_n \varkappa = -a_{n-1} - \dots - a_0 \varkappa^{-n+1}$$

implies

$$|\varkappa| < \frac{|a_{n-1}| + \dots + |a_0|}{|a_n|} \leqslant \frac{nh}{ch} = \frac{n}{c}.$$

Therefore $|\kappa| \leq \max(1, n/c) = n/c$, and the assertion (23) follows immediately.

Lemma 2. *Let* P *be a polynomial from* $\mathrm{P}_n(h)$ *and let* ω *be a real or complex number with* $\omega \in S(\kappa_1)$. *Then*

$$|\omega - \varkappa_1| < 2^n \min\left(\frac{|P(\omega)|}{|P'(\varkappa_1)|}, \left(\frac{|P(\omega)|}{|P'(\varkappa_1)|} |\varkappa_1 - \varkappa_2|\right)^{\frac{1}{2}}\right). \tag{24}$$

Proof. By (20) we have

$$\begin{aligned} |\varkappa_1 - \varkappa_i| &\leqslant |\varkappa_1 - \omega| + |\varkappa_i - \omega| \\ &\leqslant 2|\omega - \varkappa_i| \quad (i = 2, 3, \dots, n) \end{aligned} \tag{25}$$

and hence

$$|P'(\varkappa_1)| = h\prod_{i=2}^{n} |\varkappa_1 - \varkappa_i| < 2^n h \prod_{i=2}^{n} |\omega - \varkappa_i| = 2^n \frac{|P(\omega)|}{|\omega - \varkappa_1|},$$

$$|\omega - \varkappa_1| < 2^n \frac{|P(\omega)|}{|P'(\varkappa_1)|}. \tag{26}$$

Similarly, it follows from (25) that

$$\frac{|P'(\varkappa_1)|}{|\varkappa_1 - \varkappa_2|} = h\prod_{i=3}^{n} |\varkappa_1 - \varkappa_i| < 2^n h \prod_{i=3}^{n} |\omega - \varkappa_i| = 2^n \frac{|P(\omega)|}{|\omega - \varkappa_1||\omega - \varkappa_2|},$$

$$|\omega - \varkappa_1|^2 \leqslant |\omega - \varkappa_1|\,|\omega - \varkappa_2| < 2^n \frac{|P(\omega)|}{|P'(\varkappa_1)|} |\varkappa_1 - \varkappa_2|. \tag{27}$$

Clearly, the contention (24) is an immediate consequence of (26) and (27).

Lemma 3. *If $P \in \mathrm{P}_n(h)$, then for any zero κ,*

$$|P'(\varkappa)| > c(n)\,|D(P)|^{\frac{1}{2}}\,h^{-n+2} \tag{28}$$

Furthermore, if κ is not real, then

$$|P'(\varkappa)|^2 > c(n)\,|\operatorname{Im}\varkappa|\,|D(P)|^{\frac{1}{2}}\,h^{-n+3} \tag{29}$$

Proof. If we express the discriminant $D(P)$ in the form (19) we obtain

$$|D(P)| = h^{2n-2}\prod_{i\neq j}|\varkappa_i-\varkappa_j| = h^2\prod_{i\neq 1}|\varkappa_1-\varkappa_i|^2\,h^{2n-4}\prod_{\substack{i\neq j\\ i,j\neq 1}}|\varkappa_i-\varkappa_j|.$$

By Lemma 1,

$$|\varkappa_i-\varkappa_j|\leqslant 2n \quad (i,\ j=1,\ 2,\ \ldots,\ n). \tag{30}$$

Therefore,

$$|D(P)| < (2n)^{n^2}\,|P'(\varkappa_1)|^2\,h^{2n-4}$$

Since we may obviously choose for κ_1 any root of the polynomial $P(x)$, this implies the assertion (28). We wish to remark that (28) also holds in the case where the coefficients of $P(x)$ are complex.

If κ_1 is a complex zero of the polynomial $P(x)$, then let the zero κ_2 be the complex conjugate of κ_1. In analogy with the previous consideration we have

$$|D(P)| = |P'(\varkappa_1)|^2\,|P'(\varkappa_2)|^2\,h^{2n-6}\,|\varkappa_1-\varkappa_2|^{-2}\prod_{\substack{i\neq j\\ i,j\neq 1,2}}|\varkappa_i-\varkappa_j|.$$

Since $|\kappa_1-\kappa_2| = 2|\operatorname{Im}\kappa_1|$, we obtain, again applying (30), the inequality

$$|D(P)| < (2n)^{n^2}\,|P'(\varkappa_1)|^4\,|\operatorname{Im}\varkappa_1|^{-2}\,h^{2n-6}.$$

Lemma 4. *Under the conditions of Lemma 2,*

$$|\omega-\varkappa_1| < c(n)\,h^{n-2}\,|D(P)|^{-\frac{1}{2}}\,|P(\omega)|; \tag{31}$$

furthermore, if κ_1 is a nonreal zero, then also

$$|\omega-\varkappa_1|^2 < c(n)\,|\operatorname{Im}\varkappa_1|^{-1}\,h^{n-3}\,|D(P)|^{-\frac{1}{2}}\,|P(\omega)|^2. \tag{32}$$

Proof. We apply Lemmas 2 and 3. Obviously, the inequality (31) remains valid for polynomials $P(x)$ with complex coefficients.

Lemma 5. *Under the conditions of Lemma 2, let*

$$|P(\omega)| < h^{-w},\ w > 0.$$

Then

$$|\omega - \varkappa_1| < c(n) h^{-1-w_1} |D(P)|^{-\frac{1}{6}}. \tag{33}$$

where $w_1 = (2w - n)/3$ $(n \geq 3)$. *If* ω *is a complex number with* $|\mathrm{Im}\,\omega| > h^{-1/2n}$, *then*

$$|\omega - \varkappa_1|^2 < c(n) |\mathrm{Im}\,\omega|^{-2n} h^{-\frac{5}{3} - w_2} |D(P)|^{-\frac{1}{6}}, \tag{34}$$

where $w_2 \geq (4w - n)/3$ $(n \geq 4)$.

Proof. Let $\rho > 0$ be any real number. The equation $|D(P)| = |P'(\kappa_1)|^2 |D(P_1)|$ with $P_1(x) = P(x)(x - \kappa_1)^{-1}$, as was shown in the proof of Lemma 3, implies the inequality $|P'(\kappa_1)|^2 > |D(P)| h^{-2n+4+2\rho}$ provided that $|D(P_1)| < h^{2n-4-2\rho}$. Consequently, we obtain from (26)

$$|\omega - \varkappa_1| < c(n) \frac{|P(\omega)|}{|P'(\varkappa_1)|} < c(n) h^{-w+n-2-\rho} |D(P)|^{-\frac{1}{2}}.$$

On the other hand, if $|D(P_1)| \geq h^{2n-4-2\rho}$, it follows by applying Lemma 4 to the polynomial $P_1(x)$ that

$$|\omega - \varkappa_2| < c(n) h^{n-3} |D(P_1)|^{-\frac{1}{2}} |P_1(\omega)| \leq c(n) \frac{h^{-w-1+\rho}}{|\omega - \varkappa_1|}$$

where κ_2 is a zero of $P_1(x)$ whose distance from ω is minimal.

Thus we have

$$|\omega - \varkappa_1| < c(n) \max \left(h^{-w+n-2-\rho} |D(P)|^{-\frac{1}{2}},\ h^{-\frac{w-\rho+1}{2}} \right)$$

for any $\rho > 0$. We choose ρ such that the two quantities within the parentheses become equal to each other, and this yields the assertion (33).

If ω is a complex (i.e. nonreal) number, then κ_1 must be a complex zero of the polynomial $P(x)$; otherwise we would have

$$|\omega - \varkappa_i| \geqslant |\omega - \varkappa_1| \geqslant |\mathrm{Im}\,\omega| \quad (i = 1, 2, \ldots, n).$$

This would imply $h |\mathrm{Im}\,\omega|^n \leq |P(\omega)| < h^{-w}$, which is impossible. Now let $P_2(x) = P(x)(x - \kappa_1)^{-1}(x - \bar{\kappa}_1)^{-1}$; then

$$|D(P)| = |P'(\varkappa_1)|^2 |P'(\bar{\varkappa}_1)|^2 \frac{|D(P_2)|}{|\varkappa_1 - \bar{\varkappa}_1|^2} = |P'(\varkappa_1)|^4 \frac{|D(P_2)|}{|2\,\mathrm{Im}\,\varkappa_1|^2}.$$

If $|D(P_2)| < h^{2n-6-2\rho}$, this implies

$$|P'(\varkappa_1)|^4 > c(n) |D(P)| h^{-2n+6+2\rho} |\mathrm{Im}\,\varkappa_1|^2.$$

Consequently, it follows from Lemma 2 that

$$|\omega-\varkappa_1|^2<c(n)\frac{|P(\omega)|^2}{|P'(\varkappa_1)|^2}<c(n)|\mathrm{Im}\,\omega|^{-1}h^{-2w-\rho+n-3}|D(P)|^{-\frac{1}{2}}.$$

In this connection we observe that for $h>c(n)$ we have $|\mathrm{Im}\,\kappa_1|>\frac{1}{2}|\mathrm{Im}\,\omega|$, since the assumed inequality $|P(\omega)|<h^{-w}$ implies $|\omega-\kappa_1|<h^{-1/n(1+w)}$.

On the other hand, if $|D(P_2)|\geq h^{2n-6-2\rho}$, then we apply Lemma 4 to the polynomial $P_2(x)$, and thus we obtain

$$\begin{aligned}|\omega-\varkappa_2|^2&<c(n)\,|\mathrm{Im}\,\varkappa_2|^{-1}h^{n-5}|D(P_2)|^{-\frac{1}{2}}|P_2(\omega)|^2\\&<c(n)\,|\mathrm{Im}\,\varkappa_2|^{-1}h^{-2w+\rho-2}|\omega-\varkappa_1|^{-2}|\omega-\bar{\varkappa}_1|^{-2}\\&<c(n)\,|\mathrm{Im}\,\omega|^{-3}h^{-2w+\rho-2}|\omega-\varkappa_1|^{-2},\end{aligned}$$

where κ_2 is a zero of the polynomial $P_2(x)$ with minimal distance from ω. Here we may assume that $|\omega-\kappa_2|<\frac{1}{2}|\mathrm{Im}\,\omega|$; otherwise we would have

$$\frac{1}{2}|\mathrm{Im}\,\omega|\leqslant|\omega-\varkappa_2|\leqslant|\omega-\varkappa_i|\leqslant|\varkappa_1-\varkappa_i|+h^{-\frac{1}{n}(1+w)},$$

for any root $\kappa_i\neq\kappa_1,\bar{\kappa}_1$. Therefore,

$$|\varkappa_1-\varkappa_i|>\frac{1}{4}|\mathrm{Im}\,\omega|\quad(i=2,3,\ldots,n),$$

if $h>c(n)$, and hence

$$|P'(\varkappa_1)|>4^{-n}|\mathrm{Im}\,\omega|^{n-1}h,\ |\omega-\varkappa_1|<c(n)\,|\mathrm{Im}\,\omega|^{-n+1}h^{-1-w},$$

which is not weaker than the assertion (34) of the lemma.

Hence

$$|\omega-\varkappa_1|^4\leqslant|\omega-\varkappa_1|^2|\omega-\varkappa_2|^2<c(n)\,|\mathrm{Im}\,\omega|^{-3}h^{-2w+\rho-2}.$$

In view of the previous remarks this implies

$$|\omega-\varkappa_1|^2<c(n)\,|\mathrm{Im}\,\omega|^{-2n}\max\left(h^{-2w-\rho+n-3}|D(P)|^{-\frac{1}{2}},\ h^{-w+\frac{\rho}{2}-1}\right).$$

By again choosing for ρ the value which will render the two quantities within the parentheses equal, we obtain (34).

These statements are true for $h(P)>c(n)$, but in the opposite case the assertions of the lemma are trivial.

Lemma 6. *Under the conditions of Lemma 2 let $P(x)$ be a polynomial with integer coefficients, $|P(\omega)|<h^{-w}$, where*

$w\geq n-1\geq 2$ *if ω is real*

$w\geq n/2-1\geq 1$ *if ω is a complex, nonreal number.*

Furthermore, let κ_2 *be a zero of* $P(x)$ *with minimal distance from* κ_1. *Then*

$$|\omega-\kappa_1| > \begin{cases} c(n)\,|P(\omega)| : |P'(\kappa_1)|, & \text{if } |\omega-\kappa_1| \leq 2\,|\kappa_1-\kappa_2|, \\ c(n)\,\left(|P(\omega)|\,|\kappa_1-\kappa_2| : |P'(\kappa_1)|\right)^{\frac{1}{2}}, & \\ \qquad \text{if } |\omega-\kappa_1| > 2\,|\kappa_1-\kappa_2|. \end{cases}$$

Proof. If $|\omega-\kappa_1| \leq 2|\kappa_1-\kappa_2|$, then

$$|\omega-\kappa_i| \leq |\omega-\kappa_1| + |\kappa_1-\kappa_i| \leq 2\,|\kappa_1-\kappa_2| + |\kappa_1-\kappa_i| \leq 3\,|\kappa_1-\kappa_i| \quad (i=2,\,3,\,\ldots,\,n).$$

Therefore,

$$\frac{|P(\omega)|}{|\omega-\kappa_1|} = h\prod_{i=2}^{n}|\omega-\kappa_i| < 3^n h\prod_{i=2}^{n}|\kappa_1-\kappa_i| = 3^n\,|P'(\kappa_1)|,$$

and hence

$$|\omega-\kappa_1| \geq 3^{-n}\,|P(\omega)| : |P'(\kappa_1)|. \tag{35}$$

Now let us assume that $|\omega-\kappa_1| > 2|\kappa_1-\kappa_2|$. Then $|\omega-\kappa_1| \leq |\kappa_1-\kappa_i|$ $(i = 3, 4, \cdots, n)$ if $h > c(n)$. Indeed, first let ω be a real number such that, say, $|\omega-\kappa_1| > |\kappa_1-\kappa_3|$. In this case we obviously have

$$\max(|\kappa_1-\kappa_2|,\ |\kappa_1-\kappa_3|,\ |\kappa_2-\kappa_3|) \leq 3\,|\omega-\kappa_1|.$$

Furthermore it follows from Lemma 5 under the present conditions that

$$|\omega-\kappa_1| < c(n)\,h^{-\frac{1}{3}(n+1)}\,|D(P)|^{-\frac{1}{6}}.$$

Consequently,

$$\prod_{1\leq i<j\leq 3}|\kappa_i-\kappa_j| < c(n)\,h^{-n-1}, \tag{36}$$

since $|D(P)| \geq 1$. By (30) we have

$$1 \leq |D(P)| = h^{2n-2}\prod_{1\leq i<j\leq n}|\kappa_i-\gamma_j|^2$$

$$< c(n)\,h^{2n-2}\prod_{1\leq i<j\leq 3}|\kappa_i-\kappa_j|^2,\qquad \prod_{1\leq i<j\leq 3}|\kappa_i-\kappa_j| > c(n)\,h^{-n+1},$$

which contradicts (36) if $h > c(n)$.

If ω is a complex, nonreal number and $|\omega-\kappa_1| > |\kappa_1-\kappa_3|$, then we have again

$$\max(|\kappa_1-\kappa_2|,\ |\kappa_1-\kappa_3|,\ |\kappa_2-\kappa_3|) \leq 3\,|\omega-\kappa_1|,$$

and in this case Lemma 5 implies

$$|\omega - \varkappa_1| < c(n)\,|\operatorname{Im}\omega|^{-n} h^{-\frac{1}{6}(n+1)} |D(P)|^{-\frac{1}{12}}.$$

Thus,

$$\prod_{1\leqslant i<j\leqslant 3} |\varkappa_i - \varkappa_j| < c(n)\,|\operatorname{Im}\omega|^{-3n} h^{-\frac{1}{2}(n+1)} \tag{37}$$

Obviously, κ_1 and κ_2 must be nonreal numbers. Therefore,

$$1 \leqslant |D(P)| = h^{2n-2} \prod_{1\leqslant i<j\leqslant n} |\varkappa_i - \varkappa_j|^2$$

$$< c(n) h^{2n-2} \prod_{1\leqslant i<j\leqslant 3} |\varkappa_i - \varkappa_j|^2 \prod_{1\leqslant i<j\leqslant 3} |\bar\varkappa_i - \bar\varkappa_j|^2$$

$$= c(n) h^{2n-2} \prod_{1\leqslant i<j\leqslant 3} |\varkappa_i - \varkappa_j|^4,$$

since to each difference $\kappa_i - \kappa_j$ $(1 \leq i < j \leq 3)$ there corresponds the difference $\bar\kappa_i - \bar\kappa_j$ of the complex conjugate roots, and under the present conditions we have either $\kappa_i \neq \bar\kappa_i$ or $\kappa_j \neq \bar\kappa_j$. Therefore,

$$\prod_{1\leqslant i<j\leqslant 3} |\varkappa_i - \varkappa_j| > c(n) h^{-\frac{1}{2}(n-1)}.$$

This inequality is incompatible with (37) if $h > c(n)|\operatorname{Im}\omega|^{-3n}$. Inasmuch as $|\operatorname{Im}\omega| > h^{-1/4n}$, we have $|\operatorname{Im}\omega|^{-3n} < h^{3/4}$. Consequently, we may conclude for $h > c(n)$ that

$$|\omega - \varkappa_1| \leqslant |\varkappa_1 - \varkappa_i| \quad (i = 3,\ 4,\ \ldots,\ n), \tag{38}$$

if $|\omega - \kappa_1| > 2|\kappa_1 - \kappa_2|$. From (38) we obtain

$$|\omega - \varkappa_i| \leqslant |\omega - \varkappa_1| + |\varkappa_1 - \varkappa_i| + |\varkappa_1 - \varkappa_i|$$
$$= 2|\varkappa_1 - \varkappa_i| \quad (i = 3,\ 4,\ \ldots,\ n).$$

Thus,

$$\prod_{i=3}^{n} |\omega - \varkappa_i| \leqslant 2^{n-2} \prod_{i=3}^{n} |\varkappa_1 - \varkappa_i|,$$

$$|P(\omega)| \leqslant 2^{n-2} |\omega - \varkappa_1|\,|\omega - \varkappa_2| \frac{|P'(\varkappa_1)|}{|\varkappa_1 - \varkappa_2|}$$

$$< 2^{n-2} \cdot \frac{3}{2} |\omega - \varkappa_1|^2 \frac{|P'(\varkappa_1)|}{|\varkappa_1 - \varkappa_2|}$$

Here we have used the fact that $|\omega-\kappa_2| \le |\omega-\kappa_1| + |\kappa_1-\kappa_2| < (3/2)\,|\omega-\kappa_1|$. Hence we obtain

$$|\omega-\varkappa_1|^2 > 2^{-n}\,\frac{|P(\omega)|}{|P'(\varkappa_1)|}\,|\varkappa_1-\varkappa_2|. \tag{39}$$

The assertion of the lemma follows from the inequalities (35) and (39).

Lemma 7. *The polynomial* (16) *satisfies the inequality*

$$\max_{(l)} |P(l)| \geqslant c(n) \max_{(i)} |a_i|,$$

where l and i run through the values $0, 1, \cdots, n$.

Proof. We consider Lagrange's interpolation polynomial for the points $x = 0, 1, \cdots, n$, i.e. the polynomial

$$P(x) = \sum_{l=0}^{n} P(l)\,\frac{A(x)}{A'(l)(x-l)}, \quad A(x) = x(x-1)\ldots(x-n).$$

By comparing coefficients of identical powers of x we obtain

$$a_i = c_{i0}P(0) + c_{i1}P(1) + \ldots + c_{in}P(n), \tag{40}$$

where the numbers c_{ij} satisfy the inequalities

$$|c_{ij}| \leqslant c(n) \quad (i, j = 0, 1, \ldots, n). \tag{41}$$

The relations (40) and (41) imply, for arbitrary i, $0 \le i \le n$, the inequality

$$|a_i| \leqslant c(n)\,(|P(0)| + |P(1)| + \ldots + |P(n)|) \leqslant c(n) \max_{(l)} |P(l)|$$

from which the assertion of the lemma follows.

Lemma 8. *If $P_1, P_2, \cdots, P_k$ are given polynomials, then there exists a constant $c > 0$, depending only on their degrees, such that*

$$h(P_1P_2\ldots P_k) > c\,h(P_1)\,h(P_2)\ldots h(P_k).$$

Proof. It suffices to prove the lemma for pairs P_1, P_2 of polynomials, since the general case follows by induction on k. Let $P = P_1P_2$, $h = h(P)$, $h_1 = h(P_1)$, $h_2 = h(P_2)$. Then we have, using (17) and (18),

$$P_{(l)} = P_{1(l)}P_{2(l)}, \quad \overline{P} = \overline{P}_1\overline{P}_2.$$

If we denote the degree of P by n we may, applying Lemma 7, choose l in such a way as to make

$$|P(l)| = \max_{(l')} |P(l')| > c(n)\,h, \tag{42}$$

where l' runs through the values $0, 1, \cdots, n$. Now we consider the polynomial

$$Q = \overline{P}_{(l)} = \overline{P}_{1(l)} \overline{P}_{2(l)}. \tag{43}$$

It follows from (42) and Lemma 1 that all roots of the polynomial Q are bounded by a constant $c(n)$, since the absolute value of the leading coefficient of the polynomial Q is equal to $|P(l)|$. Consequently, the lemma holds for the product (43) because the heights of the polynomials Q, $\overline{P}_{1(l)}$ and $\overline{P}_{2(l)}$ are of the same order of magnitude as their leading coefficients. The heights of the polynomials P and $\overline{P}_{(l)}$, P_1 and $\overline{P}_{1(l)}$, P_2 and $\overline{P}_{2(l)}$ are, respectively, of the same order of magnitude.

A refined version of this lemma, with an explicit value of the constant c, is contained in the monograph [14] by A. O. Gel'fond.

§3. LEMMAS ON MEASURABLE SETS

Lemma 9. *Let Δ be a measurable, linear point set, $\epsilon > 0$ a real number and* meas $\Delta < \epsilon$; *furthermore, let $\Lambda = \bigcup_{i=1}^{\infty} \lambda_i$ be a system of bounded intervals λ_i satisfying the conditions*

$$\operatorname{meas}(\lambda_i \cap \Delta) \geqslant \frac{1}{2} \operatorname{meas} \lambda_i \quad (i = 1, 2, \ldots).$$

Then meas $\Lambda < 4\epsilon$.

The proof is completely analogous to the proof of the following lemma.

Lemma 10. *Let Δ be a measurable planar set, $\epsilon > 0$ a real number,* meas $\Delta < \epsilon$, *and let $\Lambda = \bigcup_{i=1}^{\infty} \lambda_i$ be a system of bounded domains λ_i satisfying the conditions*

$$\operatorname{meas}(\lambda_i \cap \Delta) \geqslant \frac{1}{2} \operatorname{meas} \lambda_i, \quad \operatorname{meas} \lambda_i > c d_i^2 \quad (i = 1, 2, \ldots),$$

where d_i is the diameter of the domain λ_i and $c > 0$ is a constant. Then

$$\operatorname{meas} \Lambda < \left(\frac{26}{c} + 1 \right) \varepsilon.$$

Proof. Let Δ^* be a closed set such that $\Delta^* \subset \Lambda$ and meas $(\Lambda \backslash \Lambda^*) < \epsilon$. Then Λ^* is covered by the system of domains λ_i, and hence, by the Heine-Borel Theorem, there exists a finite subsystem, say $\lambda_{i_1}, \lambda_{i_2}, \cdots, \lambda_{i_m}$, which also covers Λ^*; in other words, the set $T = \bigcup_{k=1}^{m} \lambda_{i_k}$ satisfies the condition $\Lambda^* \subset T$. Among these domains of the system T let λ_{i_1} have maximal diameter. Then we select from T the subsystem T_1 of those domains which have points in common with λ_{i_1}. Obviously, λ_{i_1} is contained in a circle with radius d_{i_1} while T_1 is contained in a circle with radius $2d_{i_1}$. Consequently, meas $T_1 < 4\pi d_{i_1}^2$. By assumption, $d_{i_1}^2 < c^{-1}$ meas λ_{i_1}, and hence

$$\operatorname{meas} T_1 < \frac{4\pi}{c} \operatorname{meas} \lambda_{i_1}. \tag{44}$$

Next, let λ_{i_2} be a domain with maximum diameter within the system $T \backslash T_1$. We select from $T \backslash T_1$ the subsystem T_2 consisting of those domains which have points in common with λ_{i_2}. By an argument analogous to the preceding one we obtain

$$\operatorname{meas} T_2 < \frac{4\pi}{c} \operatorname{meas} \lambda_{i_2}.$$

Repeated application of this method produces a finite sequence of domains $\lambda_{i_1}, \lambda_{i_2}, \cdots$ and a finite sequence $T_1, T_2, \cdots$ of subsets of T such that

$$T = T_1 \cup T_2 \cup \ldots, \tag{45}$$

and furthermore

$$\operatorname{meas} T_1 < \frac{4\pi}{c} \operatorname{meas} \lambda_{i_1}, \quad \operatorname{meas} T_2 < \frac{4\pi}{c} \operatorname{meas} \lambda_{i_2}, \ \ldots\ . \tag{46}$$

But since $\Lambda^* \subset T$, it follows from (45) and (46) by well-known properties of Lebesgue measure that

$$\operatorname{meas} \Lambda^* \leqslant \operatorname{meas} T \leqslant \sum_k \operatorname{meas} T_k < \frac{4\pi}{c} \sum_k \operatorname{meas} \lambda_{i_k}.$$

Therefore, since meas $\lambda_{i_k} \leq 2$ meas $(\lambda_{i_k} \cap \Delta)$ by assumption and since the domains λ_{i_k} are mutually disjoint, this implies

$$\sum_k \operatorname{meas} \lambda_{i_k} \leqslant 2 \sum_k \operatorname{meas} (\lambda_{i_k} \cap \Delta) = 2 \operatorname{meas} \left\{ \sum_k (\lambda_{i_k} \cap \Delta) \right\}$$

$$= 2 \operatorname{meas} \left\{ \left(\sum_k \lambda_{i_k} \right) \cap \Delta \right\} \leqslant 2 \operatorname{meas} \Delta < 2\varepsilon.$$

Consequently,

$$\operatorname{meas} \Lambda \leqslant \operatorname{meas} \Lambda^* + \varepsilon \leqslant \left(\frac{8\pi}{c} + 1 \right) \varepsilon < \left(\frac{26}{c} + 1 \right) \varepsilon.$$

Lemma 11. a) *If* M_1 *is a measurable set of real numbers such that, for any* $\omega \in M_1$ *and for any rational number* r, *the number* $\omega + r$ *also belongs to* M_1, *then either* M_1 *or its complementary set contains almost all real numbers.*

b) *If* M_2 *is a measurable set of complex numbers such that, for any* $\omega \in M_2$

and for any two rational numbers a *and* r, *the number* $a\omega + r$ *also belongs to* M_2, *then either* M_2 *or its complementary set contains almost all complex numbers.*

Proof. a) If M_1 has positive measure then a well-known theorem of Lebesgue implies the existence of a point ω_0 at which the set M_1 has metric density. Then there exists, for any $\epsilon > 0$, a number $\tau_0(\epsilon)$ such that for each positive number $\tau < \tau_0(\epsilon)$, the intersection of M_1 with the interval $(\omega_0 - \tau, \omega_0 + \tau)$ has measure greater than $2\tau(1-\epsilon)$. Consequently, since the set of rational numbers r is everywhere dense on the real line, an arbitrary interval (α, β) of length $\lambda = \beta - \alpha$ will have an intersection with the set M_1 whose measure is at least $2\tau(1-\epsilon) \times ([\lambda/2\tau] - 1)$. If we then consider the limit process as $\tau \to 0$ and, afterwaras, $\epsilon \to 0$, it follows that the set $M_1 \cap (\alpha, \beta)$ has measure $\lambda = \beta - \alpha$. Thus almost all points of the interval (α, β) belong to the set M_1.

b) Let ω_0 with $\operatorname{Im} \omega_0 \neq 0$ be a point at which the set M_2 has metric density. Then there exists for any $\epsilon > 0$ a constant $\tau_0(\epsilon)$ such that, for every positive $\tau < \tau_0(\epsilon)$, the intersection of the set M_2 with the square

$$|\operatorname{Re}\omega - \operatorname{Re}\omega_0| < \tau, \quad |\operatorname{Im}\omega - \operatorname{Im}\omega_0| < \tau,$$

has measure greater than $4\tau^2(1-\epsilon)$. Since the set of points $a\omega_0 + r$ is everywhere dense in the complex plane, the intersection of the set M_2 with an arbitrary interval of the form

$$|\operatorname{Re}\omega| < \lambda, \quad |\operatorname{Im}\omega| < \lambda \tag{47}$$

is at least

$$4\tau^2(1-\varepsilon)\left(\left[\frac{\lambda}{\tau}\right] - 1\right)^2.$$

Again we consider the limit processes $\tau \to 0$ and $\epsilon \to 0$ (in this order), with the result that the measure under consideration is at least $4\lambda^2$. Hence almost all points the square (47) belong to the set M_2.

Finally we wish to state the well-known Borel-Cantelli Lemma, which plays an important part in metrical theorems of number theory, and to which we shall frequently refer later on.

Lemma 12. *Let* $A = \bigcup_{i=1}^{\infty} A_i$ *be a union of measurable linear or planar sets satisfying*

$$\sum_{i=1}^{\infty} \operatorname{meas} A_i < \infty. \tag{48}$$

Then the set of all those points which belong to infinitely many sets A_i has measure zero.

Proof. Let B be the set of points belonging to infinitely many sets A_i. Then obviously

$$B \subseteq \bigcup_{i=m}^{\infty} A_i \quad (m = 1, 2, \ldots),$$

and hence it follows from the condition (48) of the lemma that, as m tends to infinity,

$$\text{meas } B \leqslant \sum_{i=m}^{\infty} \text{meas } A_i \to 0.$$

§4. INVARIANCE OF THE PARAMETERS $w_n(\omega)$

Let $w > 0$ and let $M_n(w)$ be the set of real numbers for which there exists a constant $c > 0$, not depending on h but possibly on ω, n, or w, such that the inequality

$$|P(\omega)| < ch^{-w}, \quad h = h(P), \tag{49}$$

is satisfied by infinitely many polynomials P of degree less than or equal to n, with integer coefficients. Furthermore, let $a \neq 0$ be an integer and let us consider the numbers

$$\omega_1 = \omega + a, \quad a\omega, \quad \frac{1}{\omega}. \tag{50}$$

For each of these numbers, the analogous inequality

$$|Q(\omega_1)| < c_1 h^{-w}, \quad h = h(Q), \tag{51}$$

is also satisfied by infinitely many polynomials Q of degree not exceeding n, with integer coefficients. Indeed, the polynomials

$$Q(x) = P(x - a), \quad a^n P\left(\frac{x}{a}\right), \quad x^n P\left(\frac{1}{x}\right)$$

are solutions of (51) for these numbers in the specified order. Hence, if $\omega \in M_n(w)$, each of the numbers ω_1 defined in (50) is also an element of the set $M_n(w)$. This implies that, for any element $\omega \in M_n(w)$, we have $a\omega + b \in M_n(\omega)$ for any pair a, b of rational numbers.

Now let I be an interval on the real line. Then it follows from Lemma 11 that

$$\frac{\operatorname{meas}\{\mathbf{M}_n(w)\cap I\}}{\operatorname{meas} I} = \phi(w) = 1 \quad \text{or} \quad 0.$$

If $w \le w_1$ then $\mathbf{M}_n(w) \supseteq \mathbf{M}_n(w_1)$ and hence $\phi(w) \ge \phi(w_1)$. In other words, $\phi(w)$ is a monotonically nonincreasing function which assumes the values 0 and 1 only. Obviously, $\phi(n-1) = 1$, and Mahler's inequality (11) implies $\phi(4n+1) = 0$. Consequently, there exists a number w_n such that

$$\phi(w) = \begin{cases} 1, & \text{if} \quad w < w_n \\ 0, & \text{if} \quad w > w_n, \end{cases}$$

and therefore

$$\frac{\operatorname{meas}\{\mathbf{M}_n(w)\cap I\}}{\operatorname{meas} I} = \begin{cases} 1, & \text{if} \quad w < w_n, \\ 0, & \text{if} \quad w > w_n. \end{cases}$$

Thus we may conclude that, given any number $\epsilon > 0$, it is true for almost all points ω of the interval I that the inequality (49) has infinitely many solutions if $w = w_n - \epsilon$ but at most finitely many solutions if $w = w_n + \epsilon$. From the fact that $w_n(\omega)$ may be expressed as the supremum of the set of those numbers w for which the inequality (49) has infinitely many solutions, we may conclude that

$$w_n(\omega) = w_n \quad (n = 1, 2, \ldots) \tag{52}$$

for almost all numbers of the interval I. It is easy to see that w_n does not depend on the choice of the interval I and therefore the equation (52) is valid for almost all real numbers.

Trivially, an analogous result may be obtained in the case of complex numbers, with different values of w_n, of course. We will use the notation

$$\frac{1}{n} w_n = \Theta_n \quad (n = 1, 2, \ldots),$$

$$\frac{1}{n} w_n = \eta_n \quad (n = 2, 3, \ldots)$$

in the real and complex cases, respectively. Our result may then be worded as follows:

Remark 1. *There exist numbers* Θ_n *and* η_n *such that almost all real and almost all complex numbers, respectively, satisfy the equations*

$$\Theta_n(\omega) = \Theta_n \quad \textit{and} \quad \eta_n(\omega) = \eta_n \quad (n = 1, 2, \ldots).$$

§5. REDUCTION TO IRREDUCIBLE POLYNOMIALS

In the proof it turns out to be very important to be able to restrict the inequality (1) to the case where only irreducible polynomials (over the rational field) are admitted as solutions. We define $\tilde{w}_n(\omega)$ as the supremum of the set of those numbers w for which the inequality (1) is satisfied by infinitely many irreducible polynomials of degree less than or equal to n. Then the following statement is true:

Remark 2. *For any transcendental number* ω,

$$\tilde{w}_n(\omega) = w_n(\omega) \quad (n = 1, 2, \ldots). \tag{53}$$

Indeed, it follows from the definition of $w_n(\omega)$ that there exists an infinite sequence of polynomials P, with integer coefficients, which satisfy the inequality

$$|P(\omega)| < h^{-v_n+\varepsilon}, \quad h = h(P), \tag{54}$$

with $v_n = w_n(\omega)$, provided that $w_n(\omega) < \infty$; otherwise, v_n may be given any positive value. If the polynomial P is reducible, then Gauss's Lemma ensures the existence of a factorization of the form $P = P_1 P_2 \cdots P_k$, where the P_i are irreducible polynomials with integer coefficients. Let us suppose that $\tilde{v}_n = \tilde{w}_n(\omega) < \infty$ and let us set $h(P_i) = h_i$ $(i = 1, 2, \cdots, k)$. Then the definition of the quantities $\tilde{w}_n(\omega)$ implies the relation

$$\begin{aligned} |P(\omega)| &= |P_1(\omega)| \ldots |P_k(\omega)| \gg h_1^{-\tilde{v}_n-\varepsilon} \ldots h_k^{-\tilde{v}_n-\varepsilon} \\ &= (h_1 \ldots h_k)^{-\tilde{v}_n-\varepsilon} \gg h^{-\tilde{v}_n-\varepsilon}, \end{aligned} \tag{55}$$

since $h_1 h_2 \cdots h_k \ll h$ by Lemma 8. Comparing the inequalities (54) and (55), we see that there exists an infinite set of values h for which $h^{v_n - \epsilon} \ll h^{\tilde{v}_n + \epsilon}$, i.e. $v_n - \epsilon \leq \tilde{v}_n + \epsilon$. Thus $v_n \leq \tilde{v}_n$, since the number $\epsilon > 0$ was arbitrary. The opposite inequality $v_n \geq \tilde{v}_n$ is trivial, inasmuch as the quantity $\tilde{v}_n$ refers to a smaller set of polynomials.

In the case $\tilde{w}_n(\omega) = \infty$ the equation (53) is trivially satisfied, both sides being equal to infinity.

§6. REDUCTION TO THE POLYNOMIALS FROM P_n

The set of all irreducible polynomials P of degree less than or equal to n, with integer coefficients, will be subdivided into classes P_n^l according to the following principle: Let $l(P)$ be the smallest positive integer satisfying the condition

$$|P(l)| = \max_{l'=0,1,\cdots,n} |P(l')|. \tag{56}$$

Then it is obvious that $0 \leq l(P) \leq n$. We denote by P_n^l the set of irreducible polynomials P of degree less than or equal to n, with integer coefficients, for which $l(P) = l$ $(l = 0, 1, \cdots, n)$.

Now let I be an arbitrary interval on the real line and let w be any real number with $0 < w < w_n$, where w_n is the number defined in §4. Then the inequality (1) is, for almost all $\omega \in I$, satisfied by infinitely many polynomials of degree less than or equal to n, with integer coefficients. Let Ω_l be the set of those $\omega \in I$ for which (1) has infinitely many solutions $P \in \mathrm{P}_n^l$. Then it is easy to see that the sets Ω_l are measurable and that their union contains almost all points of the interval I; in other words meas $\bigcup_{l=0}^n \Omega_l =$ meas I. But since meas $I > 0$, there exists an index l such that meas $\Omega_l > 0$.

For this particular l, let Ω' be the set of points ω' of the form

$$\omega' = (\omega - l)^{-1}, \quad \omega \in \Omega_l.$$

Since for $\omega \in \Omega_l$ there exist infinitely many polynomials $P \in \mathrm{P}_n^l$ satisfying the inequality (1), it follows that, for each $\omega' \in \Omega'$, the inequality

$$|Q(\omega')| \ll h^{-w}, \quad h = h(Q), \tag{57}$$

is also satisfied by infinitely many polynomials, namely by the polynomials $Q(x) = \overline{P}_l(x)$. If we denote the leading coefficient of $Q(x)$ by q_n, then we have $q_n = P(l)$, and it follows from (56) by Lemma 7 that

$$|q_n| = |P(l)| \gg h(P) \gg h(Q). \tag{58}$$

Finally we apply to the set Ω' the transformation

$$\omega' \to \omega'' = (\omega' - m_0)^{-1}, \quad \Omega' \to \Omega, \tag{59}$$

where the integer m_0 is defined as follows: If $\kappa = \kappa^{(0)}$ is any root of the polynomial Q and if $\kappa^{(k)}$ is a root of the kth derivative $Q^{(k)}$, then (58) and Lemma 1 imply the relation

$$|\varkappa^{(k)}| \leqslant c(n) = c \quad (k = 0, 1, \ldots, n-1)$$

If $m > c$ then $|m - c| \leq |m - \kappa|$ and hence

$$|Q(m)| = |q_n| |m - \varkappa_1| \ldots |m - \varkappa_n| \geqslant |q_n| (m - c)^n. \tag{60}$$

Furthermore,

$$|Q^{(k)}(m)| = n(n-1) \ldots (n-k+1) |q_n'| |m - \varkappa_1^{(k)}| \\ \ldots |m - \varkappa_{n-k}^{(k)}| < n^k |q_n| (m+c)^{n-k} \quad (k = 1, 2, \ldots, n-1). \tag{61}$$

Now let m_0 be the smallest integer m satisfying $m > c$ and $(m-c)^n > n^k (m+c)^{n-k}$ $(k = 1, 2, \cdots, n-1)$. Clearly, m_0 depends on n, and it follows from (60) and (61) that

$$|Q(m_0)| > \max_{(k)} |Q^{(k)}(m_0)| \quad (k = 1, 2, \ldots, n-1). \tag{62}$$

If we replace in the inequality (57) the polynomial $Q(x)$ by $\widetilde{P}(x) = \overline{Q}_{(m_0)}(x)$ then it follows by means of (59) and (62) that for almost all points $w'' \in \Omega$ there exist infinitely many polynomials $\widetilde{P}(x) = a_0 + a_1 x + \cdots + a_n x^n$ with the property $\max\{|a_0|, |a_1|, \cdots, |a_{n-1}|\} \le |a_n| = h(\widetilde{P})$ which satisfy the inequality

$$|\hat{P}(\omega'')| \ll h^{-w}, \quad h = h(P) \tag{63}$$

But since the set Ω_l has been so chosen that meas $\Omega_l > 0$, this implies that the sets Ω' and Ω both have positive measure.

With some modifications in the notation our result may be restated as follows:

Remark 3a. *If* $0 < w < n\theta_n$ *then on the real line there exists a bounded set* Ω *with positive measure such that, for each of its points* ω, *the inequality* (1) *has infinitely many irreducible primitive polynomials* $P \in \mathbf{P}_n$ *as solutions.*

It is easy to verify that the same argument carries over to the case of complex numbers. In this case it leads to the following result:

Remark 3b. *If* $0 < w < n\eta_n$ *then in the complex plane there exists a bounded set* Ω *of positive measure such that, for each of its points* ω, *the inequality* (1) *has infinitely many irreducible primitive polynomials* $P \in \mathbf{P}_n$ *as solutions.*

Thus our problem is reduced by virtue of Remarks 3a and 3b to determining the supremum of the sets of those numbers w for which the inequality (1) has, for each number ω from certain sets Ω of real or complex numbers with positive measure, infinitely many irreducible primitive polynomials $P \in \mathbf{P}_n$ as solutions.

§7. THE SIMPLEST SPECIAL CASES OF THE CONJECTURE

In the sequel we shall have to make use of the equations

$$\Theta_2 = 1, \tag{64}$$

and

$$\eta_2 = \frac{1}{4}, \quad \eta_3 = \frac{1}{3} \tag{65}$$

which we are now going to prove. At first we shall deal with the equations (65), pertaining to the complex case.

Lemma 13. *Let* $\{\alpha_h\}$ *be an infinite sequence of positive numbers. Then there exist, for almost all complex numbers* ω, *at most finitely many integer solutions* a_0, a_1, a_2, a_3 *with* max $(|a_0|, |a_1|, |a_2|, |a_3|) \leq h$ *of the inequality*

$$|a_0 + a_1\omega + a_2\omega^2 + a_3\omega^3| < h^{-1}\alpha_h \tag{66}$$

provided that

$$\sum_{h=1}^{\infty} h^{-1}\alpha_h^2 < \infty. \tag{67}$$

Proof. Let A be the set of those numbers ω for which the inequality (66) has infinitely many solutions and let $\omega \in A$ be a fixed element. Then we select from the sequence of polynomials $P(x) = a_0 + a_1x + a_2x^2 + a_3x^3$ satisfying (66) an infinite subsequence of polynomials P_k with the property that, letting $p_k(x) = h_k^{-1}P_k(x)$,

$$\lim_{k\to\infty} p_k(x) = p(x), \tag{68}$$

where $p(x) = \beta_0 + \beta_1x + \beta_2x^2 + \beta_3x^3$ is any polynomial and convergence is defined with respect to the metric of four-dimensional euclidean space; in other words, the equation (68) is equivalent to the statement

$$\lim_{k\to 0} \max_{i=0,1,2,3} |a_i^{(k)}h_k^{-1} - \beta_i| = 0.$$

If $\kappa_1^{(k)}$, $\kappa_2^{(k)}$, and $\kappa_3^{(k)}$ are the roots of the polynomial $p_k(x)$ and if they are so numbered that

$$\min_{i=1,2,3} |\omega - \kappa_i^{(k)}| = |\omega - \kappa_1^{(k)}| \quad (k = 1, 2, \cdots),$$

then we have $|\omega - \kappa_1^{(k)}| < \frac{1}{2}|\operatorname{Im}\omega|$ for all sufficiently large k. Hence $\kappa_1^{(k)}$ is a nonreal number (assuming ω to have a nonzero imaginary part). Therefore, $\kappa_2^{(k)} = \overline{\kappa_1^{(k)}}$ and $\kappa_3^{(k)}$ is real if the three indices are suitably numbered. But since $p(\omega) = 0$ and $\max_{i=0,1,2,3} |\beta_i| = 1$ (for any k, one of the four numbers $a_i^{(k)}h_k^{-1}$, and hence at least one of the four numbers β_i, must have absolute value one), the polynomial $p'(x)$ does not vanish identically. Furthermore, ω cannot be a multiple root of the polynomial $p(x)$, and thus $p'(\omega) \neq 0$.

Obviously, (68) implies

$$\lim_{k\to\infty} p_k'(x) = \lim_{k\to\infty} \frac{1}{h_k} P'(x) = p'(x),$$

and therefore we may conclude that

$$\lim_{k\to\infty} \frac{1}{h_k} P'(\omega) = p'(\omega) \neq 0;$$

hence there exists a constant $c(\omega) \neq 0$ such that $|P_k'(\omega)| > c(\omega)\,h$.

But since

$$\frac{P'(\omega)}{P(\omega)} = \frac{1}{\omega - \varkappa_1} + \frac{1}{\omega - \varkappa_2} + \frac{1}{\omega - \varkappa_3},$$

it follows from (66) that

$$|\omega - \varkappa_1^{(k)}| \leqslant 3\frac{|P(\omega)|}{|P'(\omega)|} < c(\omega)\,h_k^{-2}\,\alpha_{h_k}.$$

Thus, if the inequality (66) has infinitely many solutions for a given ω, there are also infinitely many algebraic numbers κ of degree less than or equal to three which satisfy the inequality

$$|\omega - \varkappa| < c(\omega)\,h^{-2}\alpha_h, \quad h = h(\varkappa). \tag{69}$$

Let us denote by $\mathrm{A}(C)$ the set of those ω for which (69) has infinitely many solutions with $c(\omega) \leq C$. Then we have

$$\mathrm{A} \subseteq \bigcup_{C=1,2,\dots} \mathrm{A}(C) = \lim_{C\to\infty} \mathrm{A}(C). \tag{70}$$

We now want to show that the set $\mathrm{A}(C)$ has measure zero. Obviously this follows from Lemma 12 since the number of those numbers κ for which $h(\kappa) = H$ is at most $4(2h+1)^3$, and the outer measure of the set defined by (69) with any value $c(\omega) \leq C$ is at most

$$\sum_{h=1}^{\infty} \pi C^2 h^{-4} \alpha_h^2\, 4(2h+1)^3 \ll \sum_{h=1}^{\infty} h^{-1}\alpha_h^2 < \infty.$$

Therefore, meas $\mathrm{A}(C) = 0$, and continuity of the measure together with (70) implies that meas $\mathrm{A} = \lim_{C\to\infty} \mathrm{A}(C) = 0$.

Lemma 14. *Let m and n be integers with $1 \leq m < n$. Then there exist, for almost all complex numbers ω, at most finitely many integer solutions of the inequality*

$$|a_0 + a_1\omega^m + a_2\omega^n| < h^{1/2}\alpha_h > 0 \tag{71}$$

satisfying $\max(|a_0|, |a_1|, |a_2|) \leq h$, *provided that*

$$\sum_{h=1}^{\infty} h^{-1}\alpha_h^2 < \infty. \tag{72}$$

Proof. Let us assume that, for a given ω, the inequality (71) has infinitely many solutions, and let us choose an infinite subsequence from these solutions such that $h = \max(|a_0|, |a_1|, |a_2|) \to \infty$ and

$$\left(\frac{a_0}{h}, \frac{a_1}{h}, \frac{a_2}{h}\right) \to (\beta_0, \beta_1, \beta_2) \tag{73}$$

Furthermore, let $p(x) = \beta_0 + \beta_1 x^m + \beta_2 x^n$ which obviously implies $p(\omega) = 0$ and $\max|\beta_i| = 1$. Then $p'(x)$ does not vanish identically. If $p'(\omega) = 0$ then ω lies on one of the rays

$$\arg\omega = \frac{2\pi k}{n-m} \quad (k = 0, 1, \dots, n-m-1) \tag{74}$$

depending on whether the number $-n\beta_2/m\beta_1$ is positive or not, as one can see from the equation $p'(\omega) = m\beta_1\omega^{m-1} + n\alpha_2\beta_2\omega^{n-1}$. The total measure of the rays (74) is zero, and thus we may assume that ω does not lie on any of them and that, consequently, $p'(\omega) \neq 0$. Then the polynomials $P(x) = a_0 + a_1 x^m + a_2 x^n$ satisfy the condition $|P'(x)| > c(\omega)h$, where $h = h(P)$ and $c(\omega) > 0$, because of (73). Using (71), we now obtain from the identity

$$\frac{P'(x)}{P(x)} = \sum_{i=1}^{n} \frac{1}{x - \varkappa_i},$$

where the numbers κ_i are the zeros of the polynomial $P(x)$, the inequality

$$|\omega - \varkappa_1| \leq n\frac{|P(\omega)|}{|P'(\omega)|} < c(\omega)\, h^{-\frac{3}{2}} \alpha_h.$$

By the same argument as in the preceding lemma we can again complete the proof without difficulty by applying Lemma 12 and (72): The number of polynomials $P(x)$ with $h(P) = h$ is at most $3(2h+1)^2$, and for any such polynomial $P(x)$, the total outer measure of the set of points ω in the circle $|\omega - \kappa| < Ch^{-3/2}\alpha_h$ is less than

$$\sum_{h=1}^{\infty} \pi C^2 h^{-3} \alpha_h^2 \cdot 3(2h+1)^2 \ll \sum_{h=1}^{\infty} h^{-1}\alpha_h^2 < \infty.$$

Obviously the equations $\eta_3 = 1/3$ and $\eta_2 = 1/4$ follow from Lemmas 13 and 14.

An important feature of the cases considered in Lemma 13 and 14 lies in the fact that, if for a given ω the inequality (1) has infinitely many polynomials P as solutions (in the precise form given by (66) and (71)), then there are always infinitely many polynomials $\tilde{P}$ satisfying

$$|\tilde{P}'(\omega)| > c(\omega)h, \quad h = h(\tilde{P}), \quad c(\omega) > 0$$

among these solutions. Once this has been established, the necessary result follows immediately from Lemma 12. This argument, however, does not carry over to general polynomials of degree 4, 5, etc., and a different approach is needed in the case of real numbers. Nevertheless, the following proof is possible on the basis of all these arguments if the polynomials in (1) are subjected to certain restrictions.

Lemma 15. *For almost all complex numbers* ω *the inequality*

$$|P(\omega)| < h^{-\frac{n-1}{2}} \alpha_h \text{ with } \alpha_h > 0 \text{ and } h = h(P) \tag{75}$$

is satisfied by at most finitely many polynomials $P(x)$ *of degree* $\leq n$, *with integer coefficients and with at least one pair of nonreal zeros, provided that* $\sum_{h=1}^{\infty} h^{-1}\alpha_h^2 < \infty$.

Proof. Supposing that for a nonreal number ω the inequality (75) has infinitely many solutions, let us choose a sequence of polynomials $P(x)$ with the property

$$\lim_{h\to\infty} \frac{P(x)}{h(P)} = p(x) \tag{76}$$

from these solutions. It may easily be deduced from the condition of the lemma that ω cannot be a multiple zero of the polynomial $p(x)$. By means of (76) this implies the inequality $|P'(\omega)| > c(\omega)\,h(P)$, and the assertion follows by the same argument as used in Lemma 13.

Every polynomial satisfying the inequality (1) has a root κ with $\omega \in S(\kappa)$. Thus one may consider the system

$$\begin{cases} |P(\omega)| < h^{-w}, & h = h(P), \\ \omega \in S(\kappa), & \kappa \in P, \end{cases} \tag{77}$$

instead of (1) (the expression $\kappa \in P$ means that κ is a root of P; we shall also use this notation in the sequel). Obviously the inequality (1) may lead, for different κ's, to different systems of the form (77) involving different roots of the polynomial P. If (1) has infinitely many solutions, so does the system (77); in this case the inequality (1) is satisfied by an infinite sequence of polynomials P, and there corresponds to the system (77) an infinite sequence of pairs (P, κ). Because of Remark 2 it suffices, for a given value of $w_n(\omega)$, to consider only irreducible polynomials in the inequality (1), and hence we need only consider irreducible polynomials in (77). If the polynomial P is irreducible, the second component in the pair (P, κ) is uniquely determined by the first up to conjugates.

Conversely, the first component is uniquely determined by the second if one assumes the polynomials P to be primitive and to have positive leading coefficients. This enables us to make the transition to primitive polynomials P (of degree n) and to regard the system (77) as a "system in one unknown" , i.e. to consider as solution of (77) not the pair (P, κ) but merely the algebraic number κ. In other words, we are interested in knowing whether, for a given ω, there exist infinitely many algebraic numbers κ (of degree n) such that $\omega \in S(\kappa)$ and that the corresponding polynomial P (the minimal polynomial of κ) satisfies, at the point ω, the condition

$$|P(\omega)| < h^{-w}, \; h = h(P) = h(\kappa).$$

Now let $\rho_0 > 0$ be an arbitrary number. Then we shall say that an algebraic number of degree n belongs to the class $K^n_{\rho_0}$ if all algebraic conjugates κ' of κ with $\kappa \neq \kappa'$ satisfy the inequality

$$|\kappa - \kappa'| \geq \rho_0. \tag{78}$$

We consider the system (77) for the numbers $\kappa \in K^n_{\rho_0}$ and we observe that, if for a given ω this system has infinitely many solutions $P_k(x)$ satisfying the condition

$$\lim_{k\to\infty} \frac{P_k(x)}{h(P_k)} = p(x)$$

for some polynomial $p(x)$ (this assumption does not entail any loss of generality), then it follows from (78) that the number ω cannot be a multiple zero of $p(x)$. Again, as in the proofs of Lemmas 13 to 15, we may conclude that

$$|P_k'(\omega)| > c(\omega)\, h(P_k), \; c(\omega) > 0, \quad (k = 1, 2, \cdots).$$

Consequently there exist infinitely many solutions of the inequality

$$|\omega - \kappa| < c(\omega)\, h^{-1-w}, \quad h = h(\kappa). \tag{79}$$

But since the number of κ's of degree n with $h(\kappa) = h$ is at most $(n+1)(2h+1)^n$, the following assertion is obtained by applying Lemma 12 to the system of sets defined by (79).

Lemma 16. *For almost all real (complex) numbers ω the system* (77) *has, for given $w > n$ ($w > (n-1)/2$) and for an arbitrary $\rho_0 > 0$, at most finitely many algebraic numbers $\kappa \in K^n_{\rho_0}$ as solutions.*

In other words, we are facing a nontrivial situation only if there are infinitely many polynomials P satisfying the system (77) which have, if the height $h(P)$ is sufficiently large, in addition to $\kappa \in P$ at least one further root $\kappa' \in P$ sufficiently close to ω.

§8. THE EQUATION $\Theta_2 = 1$

First we prove that, for arbitrary $\delta > 0$, the inequality

$$P(\omega)| < h_P^{-2-\delta}, \quad h_P = h(P), \tag{80}$$

has at most finitely many irreducible polynomials $P \in \mathbf{P}_2$, with integer coefficients, as solutions. If $\kappa \in P$ and $\omega \in S(\kappa)$ then Lemma 4 implies that $|\omega - \kappa| < ch^{-2-\delta}|D(P)|^{-1/2}$. Consequently, by Lemma 12 it suffices for the proof of our assertion to show that

$$\sum_{h=1}^{\infty} h^{-2-\delta} \sum_{P \in \mathbf{P}_2(h)} |D(P)|^{-\frac{1}{2}} < \infty, \tag{81}$$

where the inner summation is to be extended only over polynomials $P \in \mathbf{P}_2(h)$ with $D(P) \neq 0$ (only such polynomials shall be considered in the sequel). We then obtain

$$\sum_{P \in \mathbf{P}_2(h)} |D(P)|^{-\frac{1}{2}} = \sum_{\max(|a_0|,|a_1|) < h} |a_1^2 - 4a_0 h|^{-\frac{1}{2}}$$

$$= \sum_{|a_1| < h} \sum_{|a_0| < h} |a_1^2 - 4a_0 h|^{-\frac{1}{2}} \leqslant \sum_{|a_1| < h} \left(1 + 2\sum_{k=1}^{h} (kh)^{-\frac{1}{2}}\right) \ll h,$$

using the fact that for fixed a_1 and variable $a_0 = 0, \pm 1, \cdots, \pm h$, the number $a_1^2 - 4a_0 h$ runs through an arithmetic progression with difference h. From this (81) follows immediately.

The equation (64) is obtained by the following argument. If $w = 2 + \delta$ is less than $w_2 = 2\Theta_2$, then Remark 3a implies the existence of a set Ω of positive measure such that, for all its points, the inequality (80) has infinitely many irreducible polynomials $P \in \mathbf{P}_2$ as solutions. However, we have shown that, for almost all ω, the inequality is solved by at most finitely many irreducible polynomials $P \in \mathbf{P}_2$. Thus $w_2 \leq 2 + \delta$ and hence $w_2 \leq 2$ since the number $\delta > 0$ was arbitrary. The reverse inequality $w_2 \geq 2$ is already known to us.

Lemma 17. *The equation $\Theta_2 = 1$ holds; that is, for almost all real numbers*

ω and arbitrary $\delta > 0$, the inequality

$$|a_0 + a_1\omega + a_2\omega^2| < h^{-2-\delta} \quad \textit{with } h = \max(|a_0|, |a_1|, |a_2|)$$

has at most finitely many integer solutions a_0, a_1, a_2.

Applications of Remarks 3a and 3b like the one we have just made will occur again in the sequel.

CHAPTER 2

THE COMPLEX CASE

§1. THE DOMAINS $\sigma_i(P)$

We now approach the general case of Mahler's conjecture, beginning with the consideration of the complex case. In the sequel the polynomials $P(x)$ shall always have integer coefficients, and they shall be irreducible, primitive and of fixed degree $n \geq 4$; the set of such polynomials satisfying the condition (21) shall be denoted by P_n. We let $w_n = n\eta_n$ $(n = 2, 3, \cdots)$, where the numbers w_n and P_n are defined as in §4 of Chapter I.

We shall consider polynomials $P \in \mathrm{P}_n$ satisfying the inequality (1) at points ω which belong to a bounded domain Ω_0 of the complex plane defined as follows: Let R and ρ with $R > 1 > \rho > 0$ be arbitrary, henceforth fixed real numbers. In the upper half of the complex plane we choose a semicircle with radius $R \geq n$ and center at the origin. From this semicircle we remove all points whose distance from the real axis is $\leq \rho$. The set of remaining points is denoted by Ω_0, i.e.

$$\Omega_0 = \{\omega;\ |\omega| \leqslant R,\ \operatorname{Im}\omega > \rho > 0\}. \tag{82}$$

Furthermore, we choose an arbitrary, fixed number $\delta > 0$ and let

$$w_0 = w_{n-1} + \delta. \tag{83}$$

Let $P \in \mathrm{P}_n$ be an arbitrary polynomial and let us denote by $\sigma(P)$ the set of all $\omega \in \Omega_0$ satisfying the inequality

$$|P(\omega)| < h_P^{-w_0}. \tag{84}$$

Then obviously $\sigma(P)$ consists of uniquely determined, pairwise disjoint (connected) domains $\sigma_i(P)$, i.e. $\sigma(P) = \Sigma\,\sigma_i(P)$. (This follows from the general theory of algebraic maps if applied to the mapping $\omega \longrightarrow P(\omega)$.) In this section we consider a fixed polynomial P, and we shall investigate properties of the domains

$\sigma_i(P)$ which will be needed in the sequel. We assume that these domains do not contain any points of the boundary of Ω_0.

Lemma 18. *Every domain $\sigma_i(P)$ contains at least one root of the polynomial P.*

Proof. For a given $w > w_0$ let us denote by $\bar{\sigma}(P, w)$ the set of those points ω for which $|P(\omega)| \leq h_P^{-w}$. Then $\bar{\sigma}(P, w)$ can be decomposed into uniquely determined, mutually disjoint domains $\bar{\sigma}_j(P, w)$ such that $\sigma_i(P)$ contains at least one domain $\bar{\sigma}_j(P, w)$. Letting w tend to infinity and observing that the intersection of all such sets $\bar{\sigma}(P, w)$ consists of the roots of the polynomial P, we obtain the assertion of the lemma.

Lemma 19. *For $h(P) > c(n, \rho)$ the domains $\sigma_i(P)$ do not contain points belonging to three different domains $S(\kappa)$, i.e. there are no three different roots κ_1, κ_2, κ_3 of a polynomial P for which the corresponding sets $S(\kappa_1)$, $S(\kappa_2)$, $S(\kappa_3)$ have a point from $\sigma_i(P)$ in common.*

Proof. Let us assume that $\sigma_i(P)$ contains points which belong to the three sets $S(\kappa_1)$, $S(\kappa_2)$, and $S(\kappa_3)$ for three different roots $\kappa_1, \kappa_2, \kappa_3$ of the polynomial, P. Obviously none of the three sets $\tau_j = \sigma_i(P) \cap S(\kappa_j)$, $(j = 1, 2, 3)$ is empty. Then one of the three sets τ_j, say τ_1, has boundary points in common with the two other sets. Let ω_0 be a point on the boundaries of both sets τ_1 and τ_2. Then $\omega_0 \in \bar{\sigma}_i(P)$ and also $\omega_0 \in S(\kappa_1) \cap S(\kappa_2)$. It follows easily from the definition of the domains $\sigma_i(P)$ that $|P(\omega_0)| \leq h_P^{-w_0}$, and furthermore the definition of the sets $S(\kappa_1)$ and $S(\kappa_2)$ implies $|\omega_0 - \kappa_1| = |\omega_0 - \kappa_2|$. By an application of the complex case of Lemma 5 one obtains

$$|\omega_0 - \kappa_1| < c(n, \rho)\, h_P^{-5/6-(4w_0-n)/6}$$

since for $h_P > c(n, \rho)$ the inequality $|\mathrm{Im}\,\omega| > h_P^{-1/2n}$ holds because of (82). Therefore, it follows from the definition of w_0 (see (83)) and from (2) that

$$\frac{4w_0 - n + 5}{6} > \frac{4w_{n-1} - n + 5}{6}$$
$$\geqslant \frac{2n - 4 - n + 5}{6} = \frac{n+1}{6}.$$

Consequently,

$$|\omega_0 - \kappa_1| = |\omega_0 - \kappa_2| < c(n, \rho)\, h_P^{-\frac{n+1}{6}},$$

and thus

$$|\kappa_1-\kappa_2|<c(n,\rho)h_P^{-\frac{n+1}{6}}. \tag{85}$$

Analogously we obtain

$$|\kappa_1-\kappa_3|<c(n,\rho)h_P^{-\frac{n+1}{6}}, \tag{86}$$

and furthermore, by means of (85) and (86),

$$|\kappa_2-\kappa_3|<c(n,\rho)h_P^{-\frac{n+1}{6}}. \tag{87}$$

Finally, an application of (85), (86), and (87) yields the inequality

$$\prod_{1\leqslant i<j\leqslant 3}|\kappa_i-\kappa_j|<c(n,\rho)h_P^{-\frac{n+1}{2}}. \tag{88}$$

On the other hand, it is easy to see that the roots κ_1, κ_2, and κ_3 must be nonreal if $h > c(n,\rho)$. Thus we obtain the following relation for the discriminant $D(P)$ of the polynomial P:

$$1\leqslant|D(P)|=h_P^{2n-2}\prod_{1\leqslant i<j\leqslant n}|\kappa_i-\kappa_j|^2$$

$$\leqslant c(n)h_P^{2n-2}\prod_{1<i<j\leqslant 3}|\kappa_i-\kappa_j|^2\prod_{1\leqslant i<j\leqslant 3}|\bar{\kappa}_i-\bar{\kappa}_j|^2$$

$$=c(n)h_P^{2n-2}\prod_{1\leqslant i<j\leqslant 3}|\kappa_i-\kappa_j|^4.$$

This inequality is incompatible with (88) for $h_P > c(n,\rho)$, which proves the assertion of Lemma 19.

Lemma 20. *If the domain* $\sigma_i(P)$ *contains points of* $S(\kappa_1)$ *and* $S(\kappa_2)$, $\kappa_1 \neq \kappa_2$, *then* κ_2 *is a zero of* P *with minimal distance from* κ_1.

Proof. By the same argument as used in the proof of the preceding lemma we again obtain the inequality (85). If κ_2 is not a zero of P which is closest to κ_1 then there exists a zero κ_3 with the property $|\kappa_1-\kappa_3|\leq|\kappa_1-\kappa_2|$, $\kappa_3\neq\kappa_1$, $\kappa_3\neq\kappa_2$. Consequently, (86) is satisfied and therefore the inequalities (87) and (88) hold. Again, as in the preceding lemma, we arrive at a contradiction.

In the sequel we shall denote the boundary of a (simply connected) domain σ by $\partial\sigma$.

Lemma 21. *If* d_i *is the diameter of the domain* $\sigma_i(P)$ *then*

$$\text{meas } \sigma_i(P) > c(n) d_i^2. \tag{89}$$

Proof. At first we remark that, if any uniquely determined domain σ contains, for a given constant $c > 0$, a point κ with the property

$$\max_{\omega \in \partial\sigma} |\omega - \kappa| < c \min_{\omega \in \partial\sigma} |\omega - \kappa|, \tag{90}$$

then the inequality $\text{meas } \sigma > \frac{1}{2}(d/c)^2$ holds, d denoting the diameter of σ. Indeed , obviously σ contains a circle about κ with radius $r = \min |\omega - \kappa|$, where the minimum is taken with respect to all points $\omega \in \partial\sigma$. Thus $\mu(\sigma) \geq \pi r^2$, and, by virtue of (90), letting $r_1 = \max_{\omega \in \partial\sigma} |\omega - \kappa| \geq d/2$, we obtain the inequality $r > r_1/c$, which then implies that $\text{meas } \sigma > \pi (d/2c)^2 > \frac{1}{2}(d/c)^2$.

Now we consider a fixed domain $\sigma_i(P)$, say $\sigma_1(P)$. According to Lemma 18, $\sigma_1(P)$ contains a zero of the polynomial P, say κ_1, and consequently also points of the domain $S(\kappa_1)$. By virtue of Lemma 19, the domain $\sigma_1(P)$ may contain points of at most one additional set $S(\kappa)$, say of $S(\kappa_2)$, with $\kappa_2 \neq \kappa_1$. We distinguish two cases:

a) $\sigma_1(P)$ does not contain points of any domain $S(\kappa)$ other than $S(\kappa_1)$.

b) $\sigma_1(P)$ contains points of $S(\kappa_2)$ with $\kappa_2 \neq \kappa_1$.

Case a). Since all points of the domain $\sigma_1(P)$ are contained in the set of $S(\kappa_1)$, we have $|\omega - \kappa_1| = \min_{i=1,\cdots,n} |\omega - \kappa_i|$.

We decompose the boundary of the domain $\sigma_1(P)$ into connected curves τ_j according to the following two possibilities:

$$|\omega - \kappa_1| \leqslant 2|\kappa_1 - \kappa_2|, \quad \omega \in \partial\sigma_1(P), \tag{91}$$
$$|\omega - \kappa_1| > 2|\kappa_1 - \kappa_2|, \quad \omega \in \partial\sigma_1(P). \tag{92}$$

Thus, a curve τ_j consists of such points $\omega \in \partial\sigma_1(P)$ for which the same one of the two inequalities (91) and (92) is satisfied, and κ_2 denotes a root of the polynomial P with minimal positive distance from κ_1. The number of these curves τ_j cannot exceed $c(n)$, since the boundary of the domain $\sigma_1(P)$ is an algebraic curve of degree $\leq c(n)$.

By means of Lemmas 2 and 6 we obtain in the case of (91)

$$c(n)\rho_1 < |\omega - \kappa_1| < c_1(n)\rho_1, \quad \rho_1 = |P(\omega)| : |P'(\kappa_1)|,$$

while in the case of (92), using the fact that $w_0 > w_{n-1} \geq n/2 - 1$, we have for $h > c(n, \rho)$,

$$c(n)\rho_2 < |\omega - \varkappa_1| < c_1(n)\rho_2,$$

$$\rho_2 = \left(|P(\omega)| \frac{|\varkappa_1 - \varkappa_2|}{|P'(\varkappa_1)|}\right)^{\frac{1}{2}}.$$

On each of these curves τ_j the inequality

$$\max_{\omega\in\tau} |\omega - \varkappa_1| < c(n) \min_{\omega\in\tau} |\omega - \varkappa_1|, \tag{93}$$

is satisfied since the equation $|P(\omega)| = h^{-w_0}$ holds for all points on the boundary of $\sigma_1(P)$.

Let us consider a pair τ_1, τ_2 of such curves with the end-point ω_0 in common. From (93) we obtain

$$\max_{\omega\in\tau_1} |\omega - \varkappa_1| < c(n) \min_{\omega\in\tau_1} |\omega - \varkappa_1| \leqslant c(n)|\omega_0 - \varkappa_1|$$
$$\leqslant c(n) \max_{\omega\in\tau_2} |\omega - \varkappa_1| < c(n) \min_{\omega\in\tau_2} |\omega - \varkappa_1|$$

and thus

$$\max_{\omega\in\tau_1} |\omega - \varkappa_1| < c(n) \min_{\omega\in\tau_2} |\omega - \varkappa_1|. \tag{94}$$

By making the transition from the pair τ_1, τ_2 to the pair τ_2, τ_3, where τ_3 is a part of the boundary of $\sigma_1(P)$ adjacent to τ_2, etc., we deduce from (94) that, for any of the curves τ_j on the boundary of $\sigma_1(P)$,

$$\max_{\omega\in\tau_1} |\omega - \varkappa_1| \leqslant c(n) \min_{\omega\in\tau_j} |\omega - \varkappa_1|. \tag{95}$$

By choosing τ_1 and τ_j to be those parts of the boundary of $\sigma_1(P)$ on which the function $|\omega - \kappa_1|$ assumes its maximum and minimum, respectively, we obtain from (95) a relationship of the form (90) for the domain $\sigma_1(P)$, and, as mentioned above, this implies the assertion of the lemma.

Case b). Assume the domain $\sigma_1(P)$ to contain points from $S(\kappa_2)$ with $\kappa_2 \neq \kappa_1$ but no points contained in sets $S(\kappa)$ with $\kappa \neq \kappa_1$ and $\kappa \neq \kappa_2$. Then a suitable line L cuts the domain $\sigma_1(P)$ into two parts which consist of the points belonging to $S(\kappa_1)$ and $S(\kappa_2)$, respectively. Since the boundary of $\sigma_1(P)$ is an algebraic curve of degree $\leq c(n)$, the line L does not intersect this boundary in more than $c(n)$ points. In this manner, the boundary is decomposed into uniquely determined, connected curves $\tau_1, \tau_2, \cdots$ and $\lambda_1, \lambda_2, \cdots$ which belong to the sets $S(\kappa_1) \cap \sigma_1(P)$ and $S(\kappa_2) \cap \sigma_1(P)$, respectively. We introduce the notation

$$m_1 = \max_{\omega\in\tau} |\omega - \varkappa_1| = m(\tau), \quad \mu_1 = \min_{\omega\in\tau} |\omega - \varkappa_1| = \mu(\tau),$$

$$m_2 = \max_{\omega\in\lambda} |\omega - \varkappa_2| = m(\lambda), \quad \mu_2 = \min_{\omega\in\lambda} |\omega - \varkappa_2| = \mu(\lambda).$$

To both types of sets τ_j and τ_i we may apply the considerations of the first case since $\tau_j \in S(\kappa_1)$ and $\lambda_i \in S(\kappa_2)$; therefore,

$$m(\tau) < c(n)\mu(\tau), \quad m(\lambda) < c(n)\mu(\lambda) \tag{96}$$

in analogy to the relation (95).

Let us consider an arbitrary curve τ_j and an adjacent curve λ_i on the boundary of $\sigma_1(P)$, and let ω_0 be a common end-point of these two curves. Then we obtain from (96)

$$m_1 < c_1\mu_1 \leqslant c_1|\omega_0 - \varkappa_1| = c_1|\omega_0 - \varkappa_2| \leqslant c_1 m_2 < c_1 c_2 \mu_2$$

(where c_1 and c_2 depend on n only), using the fact that ω_0 is on the boundary of the domains $S(\kappa_1)$ and $S(\kappa_2)$, thus having the same distance from κ_1 and κ_2. In this manner we clearly obtain the inequality $m(\tau_k) < c(n)\mu$, with $\mu = \min_{j,i}(\mu(\tau_j), \mu(\lambda_i))$, for arbitrary k. Similarly it is proved that $m(\lambda) < c(n)\mu$ and therefore

$$\max_{\tau,\lambda}(m(\tau), m(\lambda)) < c(n) \min_{\tau,\lambda}(\mu(\tau), \mu(\lambda)). \tag{97}$$

Now let α and β be two points on the boundary of $\sigma_1(P)$ with $|\alpha - \beta| = d$, where d is the diameter of $\sigma_1(P)$. If both points α and β are on the same curve τ_i, then

$$|\alpha - \beta| \leqslant |\alpha - \varkappa_1| + |\beta - \varkappa_1| \leqslant 2m(\tau). \tag{98}$$

Analogously, if α and β are on the same curve λ_j, one obtains

$$|\alpha - \beta| \le 2m(\lambda). \tag{99}$$

On the other hand, if, for example, $\alpha \in \tau_j$ and $\beta \in \lambda_i$, then $|\alpha - \beta| \le |\alpha - \kappa_1| + |\beta_1 - \kappa_2| + |\kappa_1 - \kappa_2|$. Now let ω_0 be a point on the boundary of $\sigma_1(P)$ with equal distance from κ_1 and κ_2. Then $|\kappa_1 - \kappa_2| \le |\kappa_1 - \omega_0| + |\kappa_2 - \omega_0| = 2|\kappa_1 - \omega_0| = 2|\kappa_2 - \omega_0| \le 2\min(\max_i m(\tau_i), \max_j m(\lambda_j))$. Therefore we have

$$|\alpha - \beta| \leqslant m(\tau) + m(\lambda) + 2\min(\max_{(\tau)} m(\tau), \max_{(\lambda)} m(\lambda)). \tag{100}$$

From (97), (98), (99), and (100) we may deduce

$$d \leqslant 4 \max_{(\tau,\lambda)} (m(\tau),\ m(\lambda)) < c(n)\mu. \tag{101}$$

Now we draw through κ_1 a line L_1 parallel to the line L. Then the line L_1 defines two halfplanes one of which contains L; in the other halfplane, the domain $\sigma_1(P)$ contains only points belonging to $S(\kappa_1)$, among them, in particular, all points of the semicircle in that halfplane with center κ_1 and radius $\mu_0 = \min_j \mu(\tau_j)$. Its area is equal to $\frac{1}{2}\pi\mu_0^2 \geq \frac{1}{2}\pi\mu^2 > c(n)d^2$, as one may verify by means of (101). This proves the assertion of the lemma.

§2. INESSENTIAL DOMAINS

From the set $\mathbf{P}_n$ as introduced in Chapter I, §6, we select the subset $\mathbf{P}_n(h)$ of polynomials with fixed height h. Let $P \in \mathbf{P}_n(h)$. The domains $\sigma_i(P)$ as introduced in the preceding section may contain points which belong to some of the systems $\sigma(Q)$, where the polynomial $Q \in \mathbf{P}_n(h)$ is different from P. We shall distinguish between two possibilities:

1) The set of those points of the domain $\sigma_i(P)$ which belong to some systems $\sigma(Q)$ with $Q \in \mathbf{P}_n(h)$ and $Q \neq P$ has measure less than $\frac{1}{2}$ meas $\sigma_i(P)$;

2) it does not have measure less than $\frac{1}{2}$ meas $\sigma_i(P)$.

In case 1 we call the domain $\sigma_i(P)$ essential, and in case 2, inessential. Hence the domain $\sigma_i(P)$ is essential if and only if

$$\operatorname{meas} \{\sigma_i(P) \cap \bigcup_{\substack{Q \in \mathbf{P}_n(h) \\ Q \neq P}} \sigma(Q)\} < \frac{1}{2} \operatorname{meas} \sigma_i(P). \tag{102}$$

The following result will play an important part in the further arguments.

Proposition 1. *The set of those points $\omega \in \Omega_0$ which belong to infinitely many inessential domains $\sigma_i(P)$, $P \in \mathbf{P}_n$, has measure* 0.

Proof. Suppose the point ω_0 belongs to the domain $\sigma_i(P)$, $P \in \mathbf{P}_n(h)$, and also to the domain $\sigma_j(Q)$, $Q \in \mathbf{P}_n(h)$, $P \neq Q$. Then $|P(\omega_0)| < h^{-w_0}$ and $|Q(\omega_0)| < h^{-w_0}$, where $h = h(P) = h(Q)$, and hence we obtain for the polynomial $P_1(x) = P(x) - Q(x)$ the inequality

$$|P_1(\omega_0)| < 2h^{-w_0}, \quad h(P_1) \leqslant 2h, \tag{103}$$

where obviously the polynomial $P_1(x)$ is of degree $\leq n-1$ and does not vanish identically.

Let $\Delta_n(h)$ be the set of points $\omega \in \Omega_0$ for which there exists at least one polynomial P_1, not identically 0, of degree $\leq n-1$ and of height $\leq 2h$, satisfying

$$|P_1(\omega)| < 2h^{-w_0} \text{ with } w_0 = w_{n-1} + \delta. \tag{104}$$

By the preceding argument we obtain $\omega \in \Delta_n(h)$ if $\omega \in \sigma_i(P) \cap \sigma_j(Q)$ for two different polynomials $P, Q \in \mathbf{P}_n(h)$. Hence

$$\bigcup_{\substack{Q \in \mathbf{P}_n(h) \\ Q \neq P}} \sigma_i(P) \cap \sigma(Q) \subseteq \Delta_n(h),$$

and since

$$\sigma_i(P) \cap \Big(\bigcup_{\substack{Q \in \mathbf{P}_n(h) \\ Q \neq P}} \sigma(Q)\Big) = \bigcup_{\substack{Q \in \mathbf{P}_n(h) \\ Q \neq P}} \sigma_i(P) \cap \sigma(Q) \subseteq \Delta_n(h),$$

we obviously have

$$\sigma_i(P) \cap \Big(\bigcup_{\substack{Q \in \mathbf{P}_n(h) \\ Q \neq P}} \sigma(Q)\Big) \subseteq \sigma_i(P) \cap \Delta_n(h).$$

Consequently, for an inessential domain $\sigma_i(P)$,

$$\text{meas}\,\{\sigma_i(P) \cap \Delta_n(h)\} \geqslant \frac{1}{2}\,\text{meas}\,\sigma_i(P). \tag{105}$$

Putting $\Delta_n^{h_0} = \bigcup_{h \geq h_0} \Delta_n(h)$ $(h_0 = 1, 2, \cdots)$, we obtain $\Delta_n^{h_0+1} \subseteq \Delta_n^{h_0}$ $(h_0 = 1, 2, \cdots)$, and thus we may define

$$\Delta_n = \bigcap_{h_0=1}^{\infty} \Delta_n^{h_0} = \lim \Delta_n^{h_0} \quad (h_0 \to \infty). \tag{106}$$

Obviously, Δ_n contains exactly those points $\omega \in \Omega_0$ which belong to infinitely many sets $\Delta_n(h)$, i.e. for which there exist infinitely many polynomials P_1 as solutions of the inequality (103). From (103) it follows that $|P_1(\omega)| < 2(\tfrac{1}{2} h_1)^{-w_0}$, where $h_1 = h(P_1)$. Thus, by the definition of the quantity w_{n-1}, the inequality (103) may have infinitely many solutions only for a set of measure zero. This implies that meas $\mathbf{P}_n = 0$, and hence we obtain from (106), using the continuity of the measure,

$$\lim_{h_0 \to \infty} \text{meas}\,\Delta_n^{h_0} = \text{meas}(\lim_{h_0 \to \infty} \Delta_n^{h_0}) = \text{meas}\,\Delta_n = 0.$$

Consequently, for any $\epsilon > 0$, we necessarily have

$$\text{meas}\,\Delta_n^{h_0} < \varepsilon, \quad h_0 > h_0(\varepsilon). \tag{107}$$

Finally, returning to the consideration of inessential domains $\sigma_i(P)$, we obtain by virtue of (105)

$$\operatorname{meas}(\sigma_i(P)\cap\Delta_n^{h_0})\geqslant\operatorname{meas}(\sigma_i(P)\cap\Delta_n(h))\geqslant\frac{1}{2}\operatorname{meas}\sigma_i(P), \tag{108}$$

where $h = h(P)$ and $h \geq h_0$. We denote by $\Lambda(h_0)$ the union of all inessential domains $\sigma_i(P)$ corresponding to polynomials $P \in \mathrm{P}_n$ with $h(P) \geq h_0 > h_0(\epsilon)$. Using (89), (107), and (108) we may apply Lemma 10 to the system $\Lambda = \Lambda(h)$ of domains and to the set $\Delta = \Delta_n^{h_0}$. Therefore, $\operatorname{meas}(\Lambda(h_0)) < c(n)\epsilon$ for $h_0 > h_0(\epsilon)$, and hence

$$\lim \operatorname{meas}\Lambda(h_0) = 0 \quad (h_0 \longrightarrow \infty). \tag{109}$$

If now Λ_0 is the set of points $\omega \in \Omega_0$ belonging to infinitely many inessential domains $\sigma_i(P)$, $P \in \mathrm{P}_n$, then $\Lambda_0 = \lim_{h_0 \to \infty}\Lambda(h_0)$. By virtue of (109), $\mu(\Lambda_0) = 0$, q. e. d.

In closing this section we prove a result on the number of polynomials which have essential domains.

Lemma 22. *Let λ be an arbitrary positive number and $N(h, \lambda)$ the number of polynomials $P \in \mathrm{P}_n(h)$ which have at least one essential domain $\sigma_i(P)$ satisfying* $\operatorname{meas}\sigma_i(P) \geq \lambda$. *Then*

$$N(h, \lambda) \leqslant \frac{2}{\lambda}\operatorname{meas}\Omega_0.$$

Proof. Let $\sigma_i^*(P)$ be the remainder of $\sigma_i(P)$ after canceling all points which belong to domains $\sigma_j(Q)$, $Q \neq P$, $Q \in \mathrm{P}_n(h)$. By the definition of an essential domain, $\operatorname{meas}\sigma_i^*(P) \geq \frac{1}{2}\operatorname{meas}\sigma_i(P)$, and the domains $\sigma_i^*(P)$ do not overlap. Consequently, considering only essential domains for the polynomials $P \in \mathrm{P}_n(h)$, we obtain

$$\lambda N(h, \lambda) \leqslant \sum_{P,i}\operatorname{meas}\sigma_i(P) \leqslant 2\sum_{P,i}\operatorname{meas}\sigma_i^*(P)$$

$$= 2\operatorname{meas}\Big(\sum_{P,i}\sigma_i^*(P)\Big) \leqslant 2\operatorname{meas}\Omega_0,$$

since $\sum_{P,i}\sigma_i(P)$ is a subset of Ω_0.

§3. DECOMPOSITION INTO ϵ-CLASSES

Let ϵ be an arbitrary, but fixed, positive number. We divide the roots κ of the polynomials $P \in \mathrm{P}_n$ into classes (ϵ-classes) in the following manner.

Let $\kappa = \kappa_1, \kappa_2, \cdots, \kappa_k$ be all the roots of the minimal polynomial P of κ which are located in the upper halfplane and which satisfy the condition

$$|\varkappa_1-\varkappa_2|\leqslant|\varkappa_1-\varkappa_3|\leqslant\ldots\leqslant|\varkappa_1-\varkappa_k|<1. \tag{110}$$

We assume $k\geq 2$, introduce the quantities m and ρ_i by the equations

$$m=\left[\frac{n}{\varepsilon}\right]+1,\quad |\varkappa_1-\varkappa_i|=h^{-\rho_i}\quad (i=2,\ 3,\ \ldots,\ k) \tag{111}$$

and define integers $r_2, r_3, \cdots, r_k$ satisfying the inequalities

$$\frac{r_i}{m}\leqslant\rho_i<\frac{r_i+1}{m}\qquad (i=2,\ 3,\ \ldots,\ k). \tag{112}$$

Then, obviously,

$$h^{-(r_i+1)/m}<|\varkappa_1-\varkappa_i|\leqslant h^{-r_i/m}\quad (i=2,\ 3,\ \ldots,\ k). \tag{113}$$

Thus, by (110), $\rho_2\geq\rho_3\geq\cdots\geq\rho_k>0$, and (112) implies $r_i=[m\rho_i]$; hence $r_2\geq r_3\geq\cdots$ $\cdots\geq r_k\geq 0$. In this manner we may choose for each root $\kappa=\kappa_1$ of the polynomial $P\in\mathbf{P}_n$ a vector $\mathfrak{r}=(r_2, r_3, \cdots, r_k)$ with nonnegative integer components satisfying the inequalities (113).

We do not exclude the possibility that there may exist, for given $\kappa=\kappa_1$, a system $(\kappa'_2, \kappa'_3, \cdots, \kappa'_l)$, different from the system $(\kappa_2, \kappa_3, \cdots, \kappa_k)$ but also satisfying the condition

$$|\varkappa_1-\varkappa'_2|\leqslant|\varkappa_1-\varkappa'_3|\leqslant\ldots\leqslant|\varkappa_1-\varkappa'_l|<1, \tag{114}$$

analogous to (110). But since, by definition, the polynomial P does not have other roots in the intersection of the halfplane and the circle about κ_1 with radius 1 than $\kappa_1, \kappa_2, \cdots, \kappa_k$, the system $(\kappa'_2, \kappa'_3, \cdots, \kappa'_l)$ must be a permutation of the system $(\kappa_1, \kappa_2, \cdots, \kappa_k)$. Therefore $k=l$, and it is easy to see that the equations

$$|\varkappa_1-\varkappa_i|=|\varkappa_1-\varkappa'_i|\quad (i=2,\ 3,\ \ldots,\ k) \tag{115}$$

hold. Indeed, since $\kappa'_2\in(\kappa_2, \kappa_3, \cdots, \kappa_k)$, we have, by (110), $|\kappa_1-\kappa'_2|\geq$ $|\kappa_1-\kappa_2|$. Analogously, since $\kappa_2\in(\kappa'_2, \kappa'_3, \cdots, \kappa'_l)$, we have, by (114), $|\kappa_1-\kappa_2|\geq|\kappa_1-\kappa'_2|$. Therefore, $|\kappa_1-\kappa_2|=|\kappa_1-\kappa'_2|$. If

$$|\varkappa_1-\varkappa_2|=|\varkappa_1-\varkappa_3|=\ldots=|\varkappa_1-\varkappa_r|$$
$$<|\varkappa_1-\varkappa_{r+1}|\leqslant\ldots,$$

$$|\varkappa_1 - \varkappa_2'| = |\varkappa_1 - \varkappa_3'| = \ldots = |\varkappa_1 - \varkappa_{r_1}'| < |\varkappa_1 - \varkappa_{r_1+1}'| \leqslant \ldots,$$

then clearly $r = r_1$. Now, by an analogous argument, we obtain $|\kappa_1 - \kappa_{r+1}| = |\kappa_1 - \kappa'_{r_1+1}|$, etc.

Equation (115) implies that the numbers $\rho_2, \rho_3, \cdots, \rho_k$, and accordingly the numbers $r_2, r_3, \cdots, r_k$, are uniquely determined by the root $\kappa = \kappa_1$ of the polynomial P.

We define the class $K_\epsilon(\mathbf{r}) = K(\mathbf{r})$ to consist of all algebraic numbers κ corresponding to a vector $\mathbf{r}$.

In the sequel we shall need the following relation concerning the quantities $r_2, r_3, \cdots, r_k$.

Lemma 23. *If the class* $K(\mathbf{r})$ *contains infinitely many elements, then*

$$\sum_{j=2}^{k} (j-1) \frac{r_j}{m} \leqslant \frac{n-1}{2}. \tag{116}$$

Proof. Considering the discriminant $D(P)$ of the polynomial $P \in \mathbf{P}_n$ with the root $\kappa = \kappa_1 \in K(\mathbf{r})$ and also the roots $\bar\kappa_1, \bar\kappa_2, \cdots, \bar\kappa_k$, we obtain

$$1 \leq |D(P)| = h^{2n-2} \prod_{1 \leq i < j \leq n} |\kappa_i - \kappa_j|^2$$

$$\leqslant c(n) h_P^{2n-2} \prod_{1 \leqslant i < j \leqslant k} |\varkappa_i - \varkappa_j|^2 \prod_{1 \leqslant i < j \leqslant k} |\bar\varkappa_i - \bar\varkappa_j|^2$$

$$= c(n) h_P^{2n-2} \prod_{1 \leqslant i < j \leqslant k} |\varkappa_i - \varkappa_j|^4.$$

Consequently,

$$\prod_{1 \leqslant i < j \leqslant k} |\varkappa_i - \varkappa_j| > c(n) h^{-\frac{n-1}{2}}. \tag{117}$$

On the other hand, an application of (110) and (111) yields

$$\prod_{1 \leqslant i < j \leqslant k} |\varkappa_i - \varkappa_j| < c(n) \prod_{1 < j \leqslant k} |\varkappa_1 - \varkappa_j|^{j-1} = c(n) h^{-\sum_{j=2}^{k} (j-1)\rho_j},$$

since $|\kappa_i - \kappa_j| \leq |\kappa_1 - \kappa_i| + |\kappa_1 - \kappa_j| \leq 2|\kappa_1 - \kappa_j|$.

Thus, by (117), we have

$$\sum_{j=2}^{k}(j-1)\rho_j \leqslant \frac{n-1}{2}+\frac{c(n)}{\ln h},$$

and therefore, by (112),

$$\sum_{j=2}^{k}(j-1)\frac{r_j}{m} \leqslant \frac{n-1}{2}+\frac{c(n)}{\ln h}, \quad h=h(P). \tag{118}$$

If $K(\mathfrak{r})$ contains infinitely many elements we may let $h \longrightarrow \infty$, arriving at (116).

Lemma 24. *The number of different classes* $K(\mathfrak{r})$ *is finite and does not exceed a constant* $c(n, \epsilon)$.

Proof. The inequality (118) implies $r_j \leq c(n, \epsilon)$, and since the r_j's are nonnegative integers, the contention of the lemma follows immediately.

§4. REDUCTION TO THE ROOTS OF A FIXED CLASS $K(\mathfrak{r})$

We are now able to approach our main problem of determining the least upper bound of the numbers w for which the inequality (1) has infinitely many solutions only on a set of points ω of measure zero. We shall assume that $\omega \in \Omega_0$, where Ω_0 is the set defined in §1 of this chapter. Again, as in Chapter 1, §7, we perform the transition to the system (77). This means that instead of the question of whether there exist, for given ω, infinitely many polynomials $P \in \mathbf{P}_n$ satisfying the inequality (1), we shall be dealing with the equivalent question of whether there exist, for given ω, infinitely many algebraic numbers κ such that $\omega \in S(\kappa)$ and such that the corresponding polynomial P (the minimum polynomial for κ) satisfies the conditions

$$|P(\omega)| < h^{-w} \text{ with } h = h(P) = h(\kappa), \quad P \in \mathbf{P}_n.$$

If ω is a nonreal number then the root κ in the system (77) must necessarily be nonreal if the height h is sufficiently large.

Since we are assuming $\omega \in \Omega_0$, we have $\operatorname{Im} \omega \geq \rho > 0$, and κ will be nonreal for $h > c(n, \rho)$. By Lemma 15 the system (77) has, if $w = (n-1)/2 + \eta$, $\eta > 0$, for almost all complex numbers at most finitely many solutions κ which then belong to polynomials P with at least one pair $(\kappa, \bar{\kappa})$ of conjugate complex roots. Consequently, we may assume that in (77), in addition to the pair $(\kappa, \bar{\kappa})$ there exists a pair $(\kappa', \bar{\kappa}')$ of complex conjugate roots of the polynomial P with $(\kappa \neq \kappa', \bar{\kappa}')$. By choosing $\rho_0 = \frac{1}{2}$ in Lemma 16 (where ρ is the number defined in §1 of this chapter) we arrive at the conclusion that for $w > (n-1)/2$ we may restrict ourselves to considering in (77) only those κ which satisfy the following conditions:

1) κ is a nonreal number;

2) there exists a root κ' of the polynomial P such that $|\kappa - \kappa'| < \rho/2$.

Obviously, the point κ lies in the upper halfplane and $\operatorname{Im} \kappa > \rho/2$ for $h > c(n, \rho)$; hence the point κ' also lies in the upper halfplane. Consequently, κ satisfies the conditions which we used in the preceding section in order to define ϵ-classes. Hence there exists a class $K(\mathfrak{r})$ to which κ belongs.

Rewording what we have stated, we may say that for arbitrary $w > (n-1)/2$ the system (77) has, for almost all $\omega \in \Omega_0$, among its solutions at most finitely many algebraic numbers κ not belonging (under the conditions of §3) to any ϵ-class. Thus, we may assume that in the system (77) the number κ is contained in some ϵ-class $K(\mathfrak{r})$. Since, by Lemma 23, the number of different classes $K(\mathfrak{r})$ is finite and bounded by a constant $c(n, \epsilon)$, we may consider instead of (77) the system

$$\left.\begin{aligned} &|P(\omega)| < h_P^{-w}, \\ &\omega \in S(\varkappa),\ \varkappa \in P, \\ &\varkappa \in K(\mathfrak{r}) \end{aligned}\right\} \tag{119}$$

for a fixed class $K(\mathfrak{r})$, i.e. the system (77) corresponds in a natural way to no more than $c(n, \epsilon)$ systems (119) belonging to different classes $K(\mathfrak{r})$.

In considering the system (119) we shall in the sequel distinguish two cases depending on certain properties of the classes $K(\mathfrak{r})$ and involving the inequalities

$$1)\quad \frac{r_2}{m} \leqslant \frac{n}{4} - \frac{s_1}{2},$$

$$2)\quad \frac{r_2}{m} > \frac{n}{4} - \frac{s_1}{2},$$

where $s_1 = (r_3 + \cdots + r_k)/m$ if $k \geq 3$, and $s_1 = 0$ if $k = 2$.

We call the class $K(\mathfrak{r})$ a class of the first kind if condition 1) is satisfied, and a class of the second kind if condition 2) holds. Our further investigation depends on whether a given class $K(\mathfrak{r})$ is of the first or second kind.

§5. CLASSES OF THE FIRST KIND

First of all we remark that (113) implies the following important inequalities. On the one hand,

$$|P'(\varkappa_1)| = h_P \prod_{i=2}^{n} |\varkappa_1 - \varkappa_i|$$

$$< c(n) h_P \prod_{i=2}^{k} |\varkappa_1 - \varkappa_i| < c(n) h_P^{1-s}, \tag{120}$$

where $s = (r_2 + \cdots + r_k)/m$. On the other hand, for any root κ of the polynomial P, different from $\kappa_1, \cdots, \kappa_k$, (100) and (82) imply the inequality $|\kappa_1 - \kappa| \geq \min(1, |\operatorname{Im} \kappa_1|) \geq \rho/2$ for $h > c(n, \rho)$, and thus, by (113) we have

$$|P'(\varkappa_1)| = h_P \prod_{i=2}^{n} |\varkappa_1 - \varkappa_i|$$

$$> c(n, \rho) h_P \prod_{i=2}^{k} |\varkappa_1 - \varkappa_i| > c(n, \rho) h_P^{1-s-\varepsilon}, \tag{121}$$

since, by (111), $(k-1)/m < n/m < \epsilon$.

Let us now consider the system (119), assuming the class $K(\mathfrak{r})$ to be of the first kind. We choose $\delta = 2/m + \epsilon$ with ϵ defined as in §3, and for each root $\kappa = \kappa_1$ we associate with the system (119) the domain $\sigma_i(P)$ as defined in §10. According to a result of §2 (Proposition 1) we only have to consider those polynomials P in (119) for which the domain $\sigma_1(P)$ is essential. In so doing we add to the conditions listed in (119) the requirement that the domain $\sigma_1(P)$ containing the root $\kappa = \kappa_1$ of the polynomial P be essential, i.e. we make the transition from (119) to the system

$$\left.\begin{array}{l} |P(\omega)| < h_P^{-w} \\ \omega \in S(\varkappa_1),\ \varkappa_1 \in P, \\ \varkappa_1 \in K(\mathfrak{r}), \\ \sigma_1(P) \text{ essential.} \end{array}\right\} \tag{122}$$

Let $\mathrm{M}_1(h, \mathfrak{r})$ be the set of those points $\omega \in \Omega_0$ which satisfy the conditions (122) for some κ_1 with $h(\kappa_1) = h(P) = h$, i.e. $P \in \mathrm{P}_n(h)$. Since by Lemma 2 and (121),

$$|\omega - \varkappa_1| \ll \frac{|P(\omega)|}{|P'(\varkappa_1)|} \ll h^{-w-1+s+\varepsilon}, \tag{123}$$

the set $\mathbf{M}_1(h, \mathfrak{r})$ is covered by a system of circles with centers at the different roots κ_1 satisfying $h(\kappa_1) = h$, with radii $\ll h^{-w-1+s+\epsilon}$. Hence it suffices, in order to determine the outer measure of $\mathbf{M}_1(h, \mathfrak{r})$, to estimate the number $N(h, \mathfrak{r})$ of different roots κ_1 satisfying the last two conditions of the system (122) with $h(\kappa_1) = h$. In order to obtain an upper bound for $N(h, \mathfrak{r})$ we apply Lemma 22, finding first a lower bound for the measure of the essential domains $\sigma_1(P)$.

By Lemma 19, a domain $\sigma_1(P)$ contains only points which do not belong to more than two domains $S(\kappa_i)$, say, $S(\kappa_1)$ and $S(\kappa_2)$. If the intersection $\sigma_1(P) \cap S(\kappa_2)$ is not empty, then, by Lemma 20, κ_2 is a root of P with minimal distance from κ_1. Let ω_1 be a point on the boundary of $\sigma_1(P)$ contained in $\sigma_1(P) \cap S(\kappa_1)$ and as close as possible to κ_1. Then, by Lemma 2,

$$|\omega_1 - \varkappa_1| < c(n)\left(\frac{|(P(\omega_1)|}{|P'(\varkappa_1)|}\ |\varkappa_1 - \varkappa_2|\right)^{\frac{1}{2}}.$$

Obviously, $|P(\omega_1)| = h^{-w_0}$, and therefore, by (113) and (121), we have

$$|\omega_1 - \varkappa_1| \ll h^{-\frac{1}{2}(w_0+1-s_1-\varepsilon)}. \tag{124}$$

But since $w_{n-1} \geq n/2 - 1$, it follows that

$$\frac{1}{2}(w_0 + 1 + s_1 - \varepsilon) \geqslant \frac{1}{2}\left(\frac{n}{2} - s_1 + \delta - \varepsilon\right)$$

$$> \frac{n}{4} - \frac{s_1}{2} + \frac{1}{m} \geqslant \frac{r_2 + 1}{m}$$

using the fact that κ_1 belongs to a class of the first kind. Hence the inequalities (113) and (124) imply that

$$|\omega_1 - \varkappa_1| < h^{-\frac{1}{m}(r_2+1)} \leqslant |\varkappa_1 - \varkappa_2|$$

for $h \gg 1$. Clearly, it follows from Lemma 6 that

$$|\omega_1 - \varkappa_1| \gg \frac{|P(\omega_1)|}{|P'(\varkappa_1)|} = h^{-w_0}|P'(\varkappa_1)|^{-1}, \tag{125}$$

since $|P(\omega_1)| = h^{-w_0}$. We assumed ω_1 to be a point on the boundary of the domain $\sigma_1(P)$, belonging to $\sigma_1(P) \cap S(\kappa_1)$, with smallest distance from ω_1. Consequently, for an arbitrary point ω_1 on the boundary of $\sigma_1(P)$, contained in $\sigma_1(P) \cap S(\kappa_1)$, the inequality (125) will still hold.

Obviously, the semicircle with radius $r = \min_{(\omega_1)} |\omega_1 - \kappa_1|$, $\omega_1 \in \partial\sigma_1(P) \cap S(\kappa_1)$, and with center at κ_1, embedded in the halfplane which does not contain any points of $S(\kappa_2)$, is entirely contained in the domain $\sigma_1(P)$, this domain being simply connected. If $\sigma_1(P) \cap S(\kappa_2)$ is empty then the circle with radius r about κ_1 will have the same property. By (125) we obtain

$$\operatorname{meas} \sigma_1(P) \geq \frac{1}{2} \pi r^2 \gg \frac{h^{-2w_0}}{|P'(\kappa_1)|^2},$$

and, using (120) in the computation, we have

$$\operatorname{meas} \sigma_1(P) \gg h^{-2(w_0+1-s)}. \tag{126}$$

Finally, applying Lemma 22, we may deduce from (126)

$$N(h, \mathfrak{r}) \ll h^{2(w_0+1-s)}. \tag{127}$$

Now, by (123) and (127),

$$\operatorname{meas} \mathbf{M}_1(h, \mathfrak{r}) \ll N(h, \mathfrak{r}) h^{-2(w+1-s-\varepsilon)} \ll h^{-2(w-w_0)+2\varepsilon}.$$

If $w > w_0 + \frac{1}{2} + \epsilon = w_{n-1} + \frac{1}{2} + \epsilon + \delta$ in the system (122), then

$$\operatorname{meas} \bigcup_{h=1}^{\infty} \mathbf{M}_1(h, \mathfrak{r}) \leq \sum_{h=1}^{\infty} \operatorname{meas} \mathbf{M}_1(h, \mathfrak{r}) < \infty.$$

Applying Lemma 12 to the system of sets $\mathbf{M}_1(h, \mathfrak{r})$ $(h = 1, 2, \cdots)$ we may conclude that for arbitrary $w > w_{n-1} + \frac{1}{2} + \epsilon + \delta$ the system (122) has, for almost all $\omega \in \Omega_0$, at most finitely many algebraic numbers κ as solutions.

But then, by Proposition 1, a similar result holds also with respect to the system (119). In fact, if $w > w_{n-1} + \frac{1}{2} + \epsilon + \delta = w_0 + \frac{1}{2} + \epsilon > w_0$, then the set of those points $\omega \in \Omega_0$ which satisfy the first two conditions of the system (119) is contained in the interior of $\sigma_1(P)$. Consequently, by Proposition 1, the system (119) is, for almost all $\omega \in \Omega_0$, satisfied by at most finitely many algebraic numbers $\kappa = \kappa_1$ for which the coresponding domain $\sigma_1(P)$ is inessential.

Together with the preceding remarks this leads to the following result:

Proposition 2. *If* $w > w_{n-1} + \frac{1}{2} + \epsilon + \delta$, *then for almost all* $\omega \in \Omega_0$ *the system* (119) *has among its solutions at most finitely many algebraic numbers* κ *of* *degree n contained in a class $K(\mathfrak{r})$ of the first kind.*

§6. CLASSES OF THE SECOND KIND

We now consider the system (119), assuming the class $K(\mathfrak{r})$ to be of the second kind.

By virtue of Lemma 23 we have $r_2/m + 2s_1 = r_2/m + 2(r_3/m + \cdots + r_k/m) \leq (n-1)/2$. Thus, since the class $K(\mathfrak{r})$ is of the second kind, $n/4 - s_1/2 + 2s_1 < r_2/m + 2s_1 \leq (n-1)/2$, $s_1 < n/6 - 1/3$. Furthermore, $2r_3/m + s_1 \leq 3s_1 < n/2 - 1$, and hence $r_3/m < n/4 - s_1/2 - 1/2$.

Therefore,

$$\frac{r_2}{m} > \frac{n}{4} - \frac{s_1}{2} > \frac{r_3}{m}. \tag{128}$$

Now let $\mathbf{P}'_n(2^t)$ be the subset of $\mathbf{P}_n$ consisting of those polynomials P which satisfy the condition

$$2^{t-1} \leqslant h(P) < 2^t \quad (t = 1, 2, \ldots). \tag{129}$$

We will show that for $t > c(n)$ there does not exist in the set $\mathbf{P}'_n(2^t)$ a pair P_1, P_2 of different polynomials with roots $\kappa_1^{(1)}, \kappa_1^{(2)}$, respectively, contained in the class $K(\mathfrak{r})$ under consideration and satisfying the inequality

$$|\varkappa_1^{(1)} - \varkappa_1^{(2)}| < cH^{-\frac{n}{4}+\frac{s_1}{2}}, \tag{130}$$

where $H = 2^{t-1}$ and c is a positive constant, $c > c(n)$. In fact, if we assume that there exists such a pair of polynomials, then we have, by (100) and (130),

$$|\varkappa_i^{(1)} - \varkappa_j^{(2)}| \leqslant |\varkappa_i^{(1)} - \varkappa_1^{(1)}| + |\varkappa_j^{(2)} - \varkappa_1^{(2)}| + |\varkappa_1^{(1)} - \varkappa_1^{(2)}|$$

$$\leqslant h_1^{-\frac{1}{m}r_i} + h_2^{-\frac{1}{m}r_j} + cH^{-\frac{n}{4}+\frac{s_1}{2}} \leqslant 2H^{-\frac{1}{m}r_{\max(i,j)}} + cH^{-\frac{n}{4}+\frac{s_1}{2}},$$

where $h_1 = h(P_1)$ and $h_2 = h(P_2)$ $(i, j = 1, 2, \cdots, k)$. Therefore, by (128),

$$|\varkappa_i^{(1)} - \varkappa_j^{(2)}| \leqslant \begin{cases} (c + \xi_H)H^{-\frac{n}{4}+\frac{s_1}{2}}, & \text{if } \max(i, j) \leqslant 2, \\ (2 + c)H^{-\frac{1}{m}r_{\max(i,j)}}, & \text{if } \max(i, j) \geqslant 3, \end{cases}$$

where $\xi_H \to 0$ for $H \to \infty$. Consequently, considering the resultant $R(P_1, P_2)$ of the polynomials P_1, P_2, we obtain (taking into account, as usual, the complex conjugate roots)

$$|R(P_1, P_2)| \leqslant (h_1 h_2)^n \prod_{1\leqslant i,j\leqslant n} |\varkappa_i^{(1)} - \varkappa_j^{(2)}|$$

$$\leqslant H^{2n} c(n) \prod_{1\leqslant i,j\leqslant k} |\varkappa_i^{(1)} - \varkappa_j^{(2)}|\,|\bar{\varkappa}_i^{(1)} - \bar{\varkappa}_j^{(2)}| \cdot$$

$$= c(n) H^{2n} \prod_{1\leqslant i,j\leqslant k} |\varkappa_i^{(1)} - \varkappa_j^{(2)}|^2$$

$$\leqslant c(n) H^{2n} (c+\xi_H)^8 H^{-8\left(\frac{n}{4}-\frac{s_1}{2}\right)} (2+c)^{k^2} \prod_{\max(i,j)\geqslant 3} H^{-\frac{2}{m} r_{\max(i,j)}}.$$

Since obviously

$$\frac{1}{m} \sum_{\max(i,j)\geqslant 3} r_{\max(i,j)} = \frac{1}{m} \sum_{j=3}^{k} r_j + \frac{2}{m} \sum_{j=3}^{k} (j-1) r_j \geqslant s_1 + 4s_1 = 5s_1,$$

we have

$$|R(P_1, P_2)| \leqslant c(n)(c+\xi_H)^8 (2+c)^{k^2} H^{-6s_1}.$$

Thus, if c is sufficiently small, $c = c(n)$, $H \gg 1$, the inequality (130) implies $R(P_1, P_2) = 0$. But since the polynomials P_1 and P_2 are irreducible, primitive, of the same degree n, and with positive leading coefficients, they must necessarily be identical.

Hence we may conclude that for a certain constant $c = c(n)$ the circle with center at κ_1 and radius $cH^{-n/4+s_1/2}$ does not contain a root $\kappa'_1 \in K(\mathfrak{r})$ of any polynomial from $\mathbf{P}'_n(2^t)$, different from P, with κ_1 among its roots. Comparing these circles for all $\kappa_1 \in K(\mathfrak{r})$ and taking into account that each of them has an area $\gg H^{-n/2+s_1}$, we arrive at the conclusion that the number of polynomials $P \in \mathbf{P}'_n(2^t)$ with at least one root contained in the class $K(\mathfrak{r})$ under consideration, $\ll H^{n/2-s_1}$.

By applying Lemma 2 and the inequality (121), we obtain from (119)

$$|\omega - \varkappa_1| \ll \left(\frac{|P(\omega)|}{|P'(\varkappa_1)|}\,|\varkappa_1 - \varkappa_2|\right)^{\frac{1}{2}} < h_P^{-\frac{1}{2}(w+1-s_1-\varepsilon)}$$

Consequently, for $P \in \mathbf{P}'_n(2^t)$, the relation

$$|\omega - \varkappa_1| \ll H^{-\frac{1}{2}(w+1-s_1-\varepsilon)}, \quad H = 2^{t-1}$$

holds.

Thus, the measure of the set $M_2(2^t, r)$ of all points $\omega \in \Omega_0$ satisfying (119) for a given class $K(r)$ of the second kind and with $P \in P'_n(2^t)$ will be

$$\ll H^{-w-1+s_1+\varepsilon} H^{\frac{n}{2}-s_1} = H^{-w-1+\frac{n}{2}+\varepsilon}$$

Since $H = 2^{t-1}$ we obtain for $w > n/2 - 1 + \epsilon$ the inequality

$$\text{meas} \bigcup_{t=1}^{\infty} M_2(2^t, r) \leq \sum_{t=1}^{\infty} \text{meas } M_2(2^t, r) < \infty.$$

Applying Lemma 12 to the system of the sets $M_2(2^t, r)$ $(t = 1, 2, \cdots)$ we may conclude that under our conditions the system (119) has, for almost all $\omega \in \Omega_0$, at most finitely many solutions if $w > n/2 - 1 + \epsilon$.

Proposition 3. *If $w > n/2 - 1 + \epsilon$, then for almost all points $\omega \in \Omega_0$ the system* (119) *has among its solutions at most finitely many algebraic numbers κ of degree n which belong to a class $K(r)$ of the second kind.*

§7. CONCLUSION OF THE PROOF

We may now finish the proof without difficulty.

First of all, by Propositions 2 and 3, if

$$w > \max\left(w_{n-1} + \frac{1}{2} + \varepsilon + \delta, \ \frac{n}{2} - 1 + \varepsilon\right) \tag{*}$$

then, for almost all $\omega \in \Omega_0$ and for an arbitrary class $K(r)$, the system (119) has among its solutions at most finitely many algebraic numbers κ of degree n whose minimal polynomial P belongs to P_n. But since, by the argument of §4, the system (77) may, given a number $w > (n-1)/2$, possess infinitely many solutions for a set of points $\omega \in \Omega_0$ of positive measure only with numbers κ which belong to some class $K(r)$; and since the number of different classes $K(r)$ is finite (Lemma 24), it follows by the preceding remarks that under the condition (*) the system (77) has for almost all $\omega \in \Omega_0$ at most finitely many solutions. It is easy to see that then the inequality (1) has, for almost all $\omega \in \Omega_0$, at most finitely many polynomials $P \in P_n$ as solutions if (*) holds. Finally, since $w_{n-1} \geq n/2 - 1$ $(n = 3, 4, \cdots)$ by (2), this is true for $w > w_{n-1} + ½ + \epsilon + \delta$. Since the choice of the numbers $\epsilon, \delta, \rho, R$ was arbitrary we may assert that the inequality (1) has, for almost all ω, at most finitely many polynomials $P \in P_n$ as solutions whenever $w > w_{n-1} + ½$.

However, if $w < w_n$, then Remark 3b implies the existence of a set of positive measure such that for all its points the inequality (1) has infinitely many polynomials $P \in \mathrm{P}_n$ as solutions. From this we may conclude in view of the preceding consideration that

$$w_n \leqslant w_{n-1} + \frac{1}{2} \quad (n = 4,\ 5,\ \ldots),$$

since, if $w_n > w_{n-1} + \tfrac{1}{2}$, we may select a number w satisfying $w_{n-1} + \tfrac{1}{2} < w < w_n$, and then the two statements made will contradict each other. Therefore,

$$w_n \leqslant \frac{n-3}{2} + w_3 \ (n = 4,\ 5,\ \ldots).$$

By (65), $w_3 = 1$ and thus $w_n \leq (n-1)/2$ $(n = 2, 3, \cdots)$. The opposite inequality $w_n \geq (n-1)/2$ $(n = 2, 3, \cdots)$ is already familiar to us (cf. (2)). Hence we have proved the following:

Theorem 1. *For almost all complex numbers* ω,

$$w_n(\omega) = \frac{n-1}{2} \quad (n = 2,\ 3,\ \ldots).$$

CHAPTER 3

THE REAL CASE

§1. DECOMPOSITION INTO ϵ-CLASSES

Turning now to the case of real numbers, we remark that the main line of argument is analogous to the case of complex numbers and the exposition will therefore be more condensed.

As before, let $\mathbf{P}_n$ be the set of irreducible and primitive polynomials of degree $n \geq 3$, with integer coefficients, satisfying the condition (21). We let $w_n = n\Theta_n$ $(n = 1, 2, \cdots)$, where Θ_n is defined as in Chapter 1, §4. We consider the real numbers which solve the inequality (1) and are contained in a given interval Ω_0. Again we use the notation

$$w_0 = w_{n-1} + \delta, \quad \delta > 0 \tag{131}$$

and we consider the set $\sigma(P)$ of all $\omega \in \Omega_0$ satisfying (1) for a fixed polynomial $P \in \mathbf{P}_n$, with $w = w_0$. Clearly $\sigma(P)$ is a system of mutually disjoint intervals $\sigma_i(P)$, i.e. $\sigma(P) = \sum \sigma_i(P)$, where the number of intervals $\sigma_i(P)$ does not exceed n. Let us consider a specified interval of the system $\sigma(P)$, say $\sigma_1(P)$. Possibly $\sigma_1(P)$ may contain points ω_0 which also belong to some other system $\sigma(Q)$ with $Q \neq P$, $h(Q) = h(P) = h$, and $Q \in \mathbf{P}_n(h)$, where $\mathbf{P}_n(h)$ is the set of polynomials from $\mathbf{P}_n$ which have a given height h. We shall call the interval $\sigma_1(P)$ essential or inessential according as to whether the set of such points in $\sigma_1(P)$ has measure less than ½ meas $\sigma_1(P)$ or not. Arguing as in the case of complex numbers (Chapter 2, §2), we may conclude by means of Lemma 9 that the set of numbers which are contained in an infinite sequence of inessential intervals $\sigma_i(P)$ has measure zero for arbitrary $\delta > 0$.

Now we return again to the inequality (1) for the system

$$\left.\begin{array}{l}|P(\omega)| < h_P^{-w}, \\ \omega \in S(\varkappa),\ \varkappa \in P\end{array}\right\} \tag{132}$$

and we shall introduce a partition of algebraic numbers κ into ϵ-classes. Let ω be a point in the interval $\sigma_1(P)$ satisfying $\omega \in S(\kappa_1)$ for a certain root κ_1 of the polynomial P. We number the remaining roots $\kappa_2, \kappa_3, \cdots, \kappa_n$ of the polynomial P in such a way that

$$|\varkappa_1 - \varkappa_2| \leqslant |\varkappa_1 - \varkappa_3| \leqslant \ldots \leqslant |\varkappa_1 - \varkappa_n|.$$

We let $|\kappa_1 - \kappa_i| = h^{-\rho_i}$ $(i = 2, 3, \cdots, k)$, choose an arbitrary but fixed number $\epsilon > 0$, define $m = n/\epsilon + 1$ and introduce integers r_i satisfying the inequalities

$$\begin{gathered}\frac{r_i}{m} \leqslant \rho_i < \frac{r_i + 1}{m} \quad (i = 2,\ 3,\ \ldots,\ k), \\ r_i = 0 \quad (i = k+1,\ \ldots,\ n),\end{gathered} \tag{133}$$

where k is the smallest index $i \neq 1$ for which $|\kappa_1 - \kappa_i| < 1$, if such indices exist. We may assume that indeed $|\kappa_1 - \kappa_i| < 1$ for some $i \neq 1$, since the opposite case is trivial (see Lemma 16). By (133) we have

$$h_P^{1-s-\epsilon} < |P'(\varkappa_1)| < c(n)\, h_P^{1-s}, \tag{134}$$

where $s = (r_2 + \cdots + r_n)/m$. Using the discriminant of the polynomial P we obtain

$$h^{-n+1} < c(n) \prod_{1 \leqslant i < j \leqslant n} |\varkappa_i - \varkappa_j| < c(n) \prod_{1 < j \leqslant n} |\varkappa_1 - \varkappa_j|^{j-1} =$$

$$= h^{-\sum\limits_{j=2}^{n} (j-1)\rho_j},$$

$$\sum_{j=2}^{n} (j-1)\rho_j < n - 1 + \frac{c(n)}{\ln h}.$$

We are only interested in such vectors $\mathbf{r} = (r_2, r_3, \cdots, r_n)$ for which there exist infinitely many roots κ_1 satisfying (133); hence

$$\sum_{j=2}^{n} (j-1)\frac{r_j}{m} \leqslant n - 1. \tag{135}$$

In particular, we may conclude again that there do not exist more than $c(n, \epsilon)$ different vectors $\mathbf{r}$. As before, we associate with a given class $K(\mathbf{r})$ the set of all those roots $\kappa = \kappa_1$ which satisfy (133). We call the class $K(\mathbf{r})$ a class of the first or second kind depending on which of the following two conditions it satisfies:

$$1^0 \ \frac{r_2}{m} \leqslant \frac{n - s_1}{2},$$

$$2^0 \ \frac{r_2}{m} > \frac{n - s_1}{2},$$

where $s_1 = (r_3 + \cdots + r_n)/m$.

Furthermore, we make the transition from the system (132) to the system

$$\left.\begin{array}{c} |P(\omega)| < h_P^{-w}, \\ \omega \in S(\varkappa),\ \varkappa \in P, \\ \varkappa \in K(\mathbf{r}) \end{array}\right\} \tag{136}$$

for a fixed class $K(\mathbf{r})$, considering as solutions the algebraic numbers κ which have a minimal polynomial $P \in \mathbf{P}_n$ rather than the polynomials P.

§2. CLASSES OF THE FIRST KIND

If $w \geq w_0$ in (136) then it follows from the remarks made above that we may replace (136) by the system

$$\left.\begin{array}{c} |P(\omega)| < h_P^{-w}, \\ \omega \in S(\varkappa_1),\ \varkappa \in P, \\ \varkappa_1 \in K(\mathbf{r}), \\ \sigma_1(P) \text{ essential}, \end{array}\right\} \tag{137}$$

where $\sigma_1(P)$ is an interval of the system $\sigma(P)$ which contains at least one point $\omega_0 \in S(\kappa_1)$. Obviously, by the first of the two conditions of the system (137), ω belongs to some interval $\sigma_1(P)$ of the system $\sigma(P)$. The set of those ω's which belong to infinitely many inessential intervals $\sigma_1(P)$ has measure 0.

Possibly, $\sigma_1(P)$ may contain points $\omega_1 \in S(\kappa)$ for some root κ of the polynomial P other than κ_1. Then, by Lemma 5,

$$|\omega_1 - \varkappa| \ll h_P^{-1-(2w_0-n)/3} \leqslant h_P^{-(n+1)/3},$$

and analogously, since $\omega_0 \in S(\kappa_1)$,

$$|\omega_0 - \varkappa_1| \ll h_P^{-(n+1)/3}.$$

Assuming ω_1 and ω_0 to be sufficiently close to each other, we therefore obtain $|\kappa_1 - \kappa| \ll h_P^{-(n+1)/3}$. By the same argument as in the proof of Lemma 19 we may conclude that there cannot exist three different roots $\kappa \in P$ for which the domain $S(\kappa)$ overlaps with $\sigma_1(P)$. If there are two such roots, which is not impossible, then it is easy to show that $(r_2 + 1)/m \geq (n+1)/3$, since $h^{-(r_2+1)/m} < |\kappa_1 - \kappa_2| \ll h^{-(n+1)/3}$. This implies that $\kappa = \kappa_2$ is a root of the polynomial P with smallest distance from κ_1 (see Lemma 20). If this root κ_1 is nonreal then necessarily $\kappa_2 = \bar{\kappa}_1$; conversely, if κ_2 is nonreal, then $\kappa_1 = \bar{\kappa}_2$. This means that κ_1 and κ_2 are either real or conjugate complex numbers. The truth of these assertions follows easily from the inequality (135).

Now we turn to the consideration of the system (137), assuming the class $K(\mathbf{r})$ to be of first kind.

Let $\mathbf{M}_1(h, \mathbf{r})$ be the set of those $\omega \in \Omega_0$ which satisfy the system (137) for some κ_1 with $h(\kappa_1) = h$. By Lemma 2 we have

$$|\omega - \varkappa_1| \ll |P(\omega)| : |P'(\varkappa_1)| \ll h_P^{-w-1+s+\varepsilon}, \tag{138}$$

keeping in mind (134). Consequently, the set $\mathbf{M}_1(h, \mathbf{r})$ is covered by a system of intervals of total length $\ll h^{-w-1+s+\epsilon}$, since the number $N(h, \mathbf{r})$ of intervals does not exceed the number of all $\kappa_1 \in K(\mathbf{r})$ satisfying $h(\kappa_1) = h$ for which the corresponding interval in $\sigma_1(P)$ is essential. In order to estimate the number $N(h, \mathbf{r})$ we again apply Lemma 22, with the understanding that instead of domains $\sigma_i(P)$ we are now considering intervals.

We assume at first that κ_1 is a real root of the polynomial P with $\kappa_1 < \kappa_2$. We determine a real number α with the properties

$$|P(\alpha)| = \frac{1}{2} h^{-w_0}, \quad \alpha < \varkappa_1,$$
$$|P(\omega)| < h^{-w_0}, \qquad \text{if} \quad \alpha < \omega < \varkappa_1. \tag{139}$$

Clearly the interval (α, κ_1) is contained in the interior of $\sigma_1(P)$, and since the roots nearest to α can only be κ_1 and κ_2, it must be κ_1, and hence $\alpha \in S(\kappa_1)$. In the case $\kappa_1 > \kappa_2$ it is obvious how the argument has to be modified.

Let us assume now that κ_1 is a nonreal root. We are only concerned with those intervals $\sigma_1(P)$ which contain at least one point ω_0 satisfying (1) with $w > w_0$. Let the number α be defined by the conditions

$$|P(\alpha)| = \frac{1}{2} h^{-w_0}, \quad \alpha > \omega_0,$$
$$|P(\omega)| < h^{-w_0}, \quad \text{if} \quad \omega_0 < \omega < \alpha. \tag{140}$$

It is easy to see that the interval (ω_0, α) is contained in the interior of $\sigma_1(P)$. As the two roots of P which are nearest to α we may, with equal right, consider κ_1 and κ_2, since $\kappa_2 = \bar{\kappa}_1$, $|\alpha - \kappa_1| = |\alpha - \bar{\kappa}_1|$.

Applying Lemma 2 and (134), we obtain from (139) and (140) the inequality

$$|\alpha - \varkappa_1| \ll \left(\frac{|P(\alpha)|}{|P'(\varkappa_1)|} |\varkappa_1 - \varkappa_2|\right)^{\frac{1}{2}} \ll h_P^{-\frac{1}{2}(w_0+1-s_1-\varepsilon)}$$

Since $|\kappa_1 - \kappa_2| > h^{-(r_2+1)/m}$ and since, in the case under consideration,

$$\frac{r_2+1}{m} \leqslant \frac{n-s_1}{2} + \frac{1}{m} < \frac{1}{2}(w_{n-1} + 1 + \delta - s_1 - \varepsilon) = \frac{1}{2}(w_0 + 1 - s_1 - \varepsilon),$$

it follows for $h \gg 1$, letting $\delta = 2/m + \epsilon$, that

$$|\alpha - \varkappa_1| < h_P^{-\frac{1}{m}(r_2+1)} \leqslant |\varkappa_1 - \varkappa_2|.$$

Theorefore, by Lemma 6 and (124),

$$|\alpha - \varkappa_1| \gg \frac{|P(\alpha)|}{|P'(\varkappa_1)|} \gg h^{-w_0-1+s} \tag{141}$$

If κ_1 is a nonreal root, then, by virtue of the inequalities $w > w_0$ and $|P(\omega)| < H_P^{-w}$, using Lemma 2 and (140) as well as (141), we have

$$|\omega_0 - \kappa_1| \ll |P(\omega)|: |P'(\kappa_1)| = o(|\alpha - \kappa_1|);$$

furthermore, $|\alpha - \omega_0| = |\alpha - \kappa_1|(1 + o(1))$, and hence

$$|\alpha - \omega_0| \gg h_P^{-w_0-1+s} \tag{142}$$

if $h_P \gg 1$.

In this manner, regardless of whether κ_1 is a real root or not, the interval $\sigma_1(P)$ contains a subinterval of length $\gg h^{-w_0-1+s}$. In the former case we may take this subinterval to be (α, κ_1), and in the latter case, using the inequalities (141) and (142), the interval (ω_0, α).

Applying Lemma 22 we find that $N(h, \mathbf{r}) \ll h^{w_0+1-s}$. For the measure of the set $M_1(h, \mathbf{r})$ we obtain by (138)

$$\text{meas } M_1(h, \ \mathbf{r}) \ll N(h, \ \mathbf{r}) \, h^{-w-1+s+\varepsilon} = h^{-(w-w_0)+\varepsilon}.$$

If, in the system (137), $w > w_0 + 1 + \epsilon = w_{n-1} + 1 + \epsilon + \delta$, then

$$\sum_{h=1}^{\infty} \text{meas } M_1(h, \ \mathbf{r}) < \infty.$$

Applying Lemma 12 we may conclude that for almost all points $\omega \in \Omega_0$ the system (137) has only finitely many solutions if $w > w_{n-1} + 1 + \epsilon + \delta$. But then, as we know, a similar result also holds with respect to the system (136).

§3. CLASSES OF THE SECOND KIND

We now consider the case of a class $K(\mathbf{r})$ of the second kind. The inequality (135) implies

$$\frac{r_2}{m} + 2s_1 = \frac{r_2}{m} + 2\left(\frac{r_3}{m} + \dots + \frac{r_n}{m}\right) \leqslant n - 1.$$

Hence, the class $K(\mathbf{r})$ being of the second kind, we have

$$\frac{n - s_1}{2} + 2s_1 < \frac{r_2}{m} + 2s_1 \leqslant n - 1, \quad s_1 < \frac{n-2}{3}.$$

Furthermore, $2\,r_3/m + s_1 \leq 3s_1 < n - 2$, and thus, $r_3/m < (n - s_1)/2 - 1$. Therefore,

$$\frac{r_2}{m} > \frac{n - s_1}{2} > \frac{r_3}{m}. \tag{143}$$

Again let $\mathbf{P}_n(2^t)$ be the subset of $\mathbf{P}_n$ consisting of the polynomials P satisfying (129). We shall show that, for suitable $c = c(n)$, there is no pair P_1, P_2 of different polynomials in the set $\mathbf{P}'_n(2^t)$ with roots $\kappa_1^{(1)}, \kappa_1^{(2)}$, respectively, contained in the class $K(\mathbf{r})$ under consideration and satisfying the inequality

$$|\varkappa_1^{(1)} - \varkappa_1^{(2)}| < cH^{-\frac{n-s_1}{2}}, \tag{144}$$

where $H = 2^{t-1}$. Obiviously, by virtue of (133) and (144), we have

$$|\varkappa_i^{(1)} - \varkappa_j^{(2)}| \leqslant |\varkappa_i^{(1)} - \varkappa_1^{(1)}| + |\varkappa_j^{(2)} - \varkappa_1^{(2)}| + |\varkappa_1^{(1)} - \varkappa_1^{(2)}|$$

$$\leqslant h_1^{-\frac{1}{m} r_i} + h_2^{-\frac{1}{m} r_j} + cH^{-\frac{n-s_1}{2}} \leqslant 2H^{-\frac{1}{m} r_{\max(i,j)}} + cH^{-\frac{n-s_1}{2}},$$

where $h_1 = h(P_1)$ and $h_2 = h(P_2)$ $(i, j = 1, 2, \cdots, n)$. Therefore, by (143),

$$|\varkappa_i^{(1)} - \varkappa_j^{(2)}| \leqslant \begin{cases} (c + \xi_H) H^{-\frac{n-s_1}{2}}, & \text{if } \max(i, j) \leqslant 2, \\ (2 + c) H^{-\frac{1}{m} r_{\max(i,j)}}, & \text{if } \max(i, j) \geqslant 3, \end{cases}$$

where $\xi_H \to 0$ as $H \to \infty$. Using now the resultant $R(P_1, P_2)$ of the polynomials P_1 and P_2, we obtain

$$|R(P_1, P_2)| \leqslant (h_1 h_2)^n \prod_{1 \leqslant i, j \leqslant n} |\varkappa_i^{(1)} - \varkappa_j^{(2)}|$$

$$\leqslant c(n) H^{2n} (c + \xi_H)^4 H^{-4\left(\frac{n-s_1}{2}\right)} (2 + c)^{n^2} \prod_{\max(i,j) \geqslant 3} H^{-\frac{1}{m} r_{\max(i,j)}}.$$

But since

$$\frac{1}{m} \sum_{\max(i,j) \geqslant 3} r_{\max(i,j)} = \frac{1}{m} \sum_{j=3}^{n} r_j + \frac{2}{m} \sum_{j=3}^{n} (j-1) r_j \geqslant 5s_1,$$

it follows that

$$|R(P_1, P_2)| \leqslant c(n) (c + \xi_H)^4 (2 + c)^{n^2} H^{-3s_1}.$$

Consequently, for sufficiently small $c = c(n)$ and $H >> 1$ the inequality (144) implies $R(P_1, P_2) = 0$, and therefore necessarily $P_1 = P_2$.

If we associate with each root $\kappa_1 \in K(\mathfrak{r})$ of a polynomial $P \in \mathbf{P}_n'(2^t)$ the circle with center at κ_1 and radius $cH^{-(n-s_1)/2}$ then, by the remarks made above, those circles corresponding to the roots κ_1 of different polynomials P cannot overlap. We observe that if κ_1 is a nonreal root of P the nearest root to κ_1 will be $\kappa_2 = \bar{\kappa}_1$, and we have

$$|\operatorname{Im} \varkappa_1| = \frac{1}{2} |\varkappa_1 - \varkappa_2| \leq \frac{1}{2} H^{-\frac{r_2}{m}} = \Theta_H H^{-\frac{n-s_1}{2}},$$

where $\Theta_H \to 0$ for $H \to \infty$. Hence the centers of those circles corresponding to nonreal roots κ_1 will have a distance from the real axis which becomes infinitely small in relation to their radii. With each root $\kappa \in K(\mathbf{r})$ having the minimal polynomial $P \in \mathbf{P}'_n(2^t)$ we shall associate the interval of real numbers contained in the circle corresponding to κ_1. Then we observe that these intervals do not overlap and have total length $>> H^{-(n-s_1)/2}$. Consequently, the number of polynomials $P \in \mathbf{P}'_n(2^t)$ with at least one root $\kappa \in K(\mathbf{r})$, is $<< H^{(n-s_1)/2}$.

By applying Lemma 2 and the inequalities (134) and (136) we obtain

$$|\omega - \varkappa_1| \ll \left(\frac{|P(\omega)|}{|P'(\varkappa_1)|} |\varkappa_1 - \varkappa_2| \right)^{\frac{1}{2}} \ll h_P^{-\frac{1}{2}(w+1-s_1-\varepsilon)}.$$

This means that for $P \in \mathbf{P}_n(2^t)$,

$$|\omega - \varkappa_1| \ll H^{-\frac{1}{2}(w+1-s_1-\varepsilon)}.$$

Now consider the set $\mathbf{M}_2(2^t, \mathbf{r})$, consisting of those $\omega \in \Omega_0$ for which (136) is satisfied, where the class $K(\mathbf{r})$ is of the second kind and $P \in \mathbf{P}'_n(2^t)$. We have

$$\operatorname{meas} \mathbf{M}_2(2^t, \mathbf{r}) \ll H^{-\frac{1}{2}(w+1-s_1-\varepsilon)} H^{\frac{n-s_1}{2}} = H^{-\frac{1}{2}(w+1-n-\varepsilon)}$$

Since $H = 2^{t-1}$ $(t = 1, 2, \cdots)$ we obtain for $w > n - 1 + \epsilon$ the inequality

$$\sum_{t=1}^{\infty} \operatorname{meas} \mathbf{M}_2(2^t, \mathbf{r}) < \infty.$$

Applying Lemma 12 to the system of sets $\mathbf{M}_2(2^t, \mathbf{r})$ $(t = 1, 2, \cdots)$, we conclude that for almost all $\omega \in \Omega_0$ the system (136) has at most finitely many solutions $\kappa \in K(\mathbf{r})$ of the second kind if $w > n - 1 + \epsilon$.

§4. CONCLUSION OF THE PROOF

Let us summarize: It has been demonstrated in §§2 and 3 that for $w > \max(w_{n-1} + 1 + \epsilon + \delta, n - 1 + \epsilon)$ the system (136) has, for almost all $\omega \in \Omega_0$ and for an arbitrary class $K(\mathbf{r})$, among its solutions at most finitely many algebraic numbers κ of degree n with minimum polynomials $P \in \mathbf{P}_n$. Hence the system (132) also has for almost all $\omega \in \Omega_0$ at most finitely many solutions, provided that the inequality (145) is satisfied, the number of different classes $K(\mathbf{r})$ not exceeding a constant $c(n, \epsilon)$. By virtue of (2) we know that $w_{n-1} \geq n - 1$ $(n = 2, 3, \cdots)$, and therefore, $\max(w_{n-1} + 1 + \epsilon + \delta, n - 1 + \epsilon) = w_{n-1} + 1 + \epsilon + \delta$. Consequently, we conclude that for almost all $\omega \in \Omega_0$ the inequality (1) has at most finitely many polynomials $P \in \mathbf{P}_n$ as solutions if $w > w_{n-1} + 1 + \epsilon + \delta$. Since the numbers ϵ and δ and the interval Ω_0 were arbitrary the analogous assertion holds for almost all real numbers if $w > w_{n-1} + 1$.

However, by Remark 3a (Chapter 1, §6) there exists a set Ω of positive measure such that the inequality (1) has, for each point of Ω, infinitely many polynomials $P \in \mathrm{P}_n$ as solutions, provided that $w < w_n$. In view of the previous considerations it follows that $w_n \leq w_{n-1} + 1$ $(n = 3, 4, \cdots)$. But since we have $w_2 = 2$ by (64), this implies $w_n \leq n$ $(n = 2, 3, \cdots)$. The opposite inequality $w_n \geq n$ $(n = 1, 2, \cdots)$ is well known to us (see (2)). Hence we have proved the following theorem.

Theorem. *For almost all real numbers* ω,

$$w_n(\omega) = n \quad (n = 1, 2, \cdots).$$

Part II

FIELDS WITH NON-ARCHIMEDEAN VALUATION

CHAPTER 1

BASIC FACTS

§1. INTRODUCTION

Instead of the fields of real and complex numbers which we discussed in Part 1, we shall now consider fields with non-archimedean valuation.

Before we turn to our main problem we want to remind the reader of the basic definitions and facts from the theory of fields with non-archimedean valuation (cf., for example, [78, 34, 35]).

Let K be a field in which a non-archimedean valuation is defined for all $\omega \in K$, i.e. a real-valued function which satisfies the following conditions:

$$|\omega| > 0 \quad (\omega \neq 0), \quad |0| = 0,$$

$$|\omega_1\omega_2| = |\omega_1|\ |\omega_2|,$$

$$|\omega_1 + \omega_2| \leq \max(|\omega_1|,\ |\omega_2|).$$

The set V of elements $\omega \in K$ with $|\omega| \leq 1$ forms a ring under addition and multiplication as defined in the field K; it is called the ring of integers in K. The subset P of the elements $\omega \in V$ with $|\omega| < 1$ forms a maximal ideal in V since outside P there are only units of V, i.e. elements which have inverses within the ring. Clearly P is a prime ideal since the factor ring V/P is a field; we call it the residue class field of the field K.

The set of values assumed by the function $l(\omega) = \log|\omega|$, defined for all $\omega \in K$, $\omega \neq 0$, forms an additive abelian group Γ, the valuation group of the field K. If this group is discrete, then the valuation is also called discrete. In this case Γ is a cyclic group and P is a principal ideal. Thus there exists a generating element π for the ideal P. This fact is expressed by the notation $P = (\pi)$.

A field K with valuation may be considered as a metric space in which the distance function is defined as $\rho(\omega_1, \omega_2) = |\omega_1 - \omega_2|$ for all pairs $\omega_1, \omega_2 \in K$. The topology of this metric space is called the natural topology of the field K. If all Cauchy sequences of elements in the field K have limits in K, i.e. if it is a complete metric space, then K is also called a complete field.

The elements of a complete field with discrete valuation may be expressed conveniently by infinite series.

Lemma 1. *Let K be a complete field with discrete valuation, $P = (\pi)$ the (maximal) valuation ideal, T a full residue system modulo P including 0. Then there exists, for any element $\omega \in K$, a unique expansion of the form*

$$\omega = \sum_{s=l}^{\infty} \varepsilon_s \pi^s, \ \varepsilon_l \neq 0, \ \varepsilon_s \in T \quad (s = l, l+1, \ldots). \tag{1}$$

A proof of this lemma may be found in the book by Lenskoĭ ([35], pp. 38–41) or in the book by Borevič and Šafarevič ([3], Chapter IV, §1.2).

§2. MEASURE ON A LOCALLY COMPACT FIELD

In the field K we shall introduce Haar measure, using the natural topology of the field K and assuming K to be locally compact, since otherwise the definition of such a measure would be impossible in general. Instead of the abstract construction of Haar (see [16], Chapter XI, §58) we shall define a measure in K by considering the field as a product $T_\infty = T \times T \times \cdots$ of infinitely many copies of the space T, based on the expansion (1) of Lemma 1 (see [16], Chapter VII, §38). We prefer the latter approach in view of the fact that in our case it furnishes the necessary properties of the measure without difficulty. The following lemma is the starting point for our investigation.

Lemma 2. *A non-archimedean field K is locally compact if and only if it is complete, if the valuation is discrete and if K has a finite residue class field.*

A proof may be found, for example, in [35], pp. 43–44.

Consequently, if the field K is locally compact, then the set T occurring in the expansions (1) is finite. Let q be the number of elements of T, i.e. the number of residue classes mod P. In the sequel we shall say that the valuation in K is defined by the equation $|\omega| = q^{-l}$, where l is defined by the equation (1).

The procedure we are going to follow can be found in detail (with a somewhat different approach) in a monograph by A. G. Postnikov ([50], §18). Therefore we restrict ourselves to a short presentation.

Let $\omega_0 \in K$ and $r \in \Gamma$. A circle $C(\omega_0, r)$ is defined as the set of all points $\omega \in K$ with $|\omega - \omega_0| \leq r$.

Lemma 3. *Let* $C_1 = C(\omega_1, r_1)$ *and* $C_2 = C(\omega_2, r_2)$ *be circles in* K *and assume* $C_1 \cap C_2$ *to be nonempty. Then either* $C_1 \subseteq C_2$ *or* $C_1 \supseteq C_2$.

Proof. Let $\omega_0 \in C_1 \cap C_2$. Then, if $r_1 \leq r_2$, we have

$$|\omega_1 - \omega_2| = |\omega_1 - \omega_0 - (\omega_2 - \omega_0)| \leqslant \max(|\omega_1 - \omega_0|, |\omega_2 - \omega_0|) \leqslant \max(r_1, r_2) = r_2$$

Now we obtain, for any $\omega \in C_1$,

$$|\omega - \omega_2| = |\omega - \omega_1 - (\omega_2 - \omega_1)| \leqslant \max(|\omega - \omega_1|, |\omega_2 - \omega_1|) \leqslant \max(r_1, r_2) = r_2,$$

and hence $\omega \in C_2$. Thus $C_1 \subseteq C_2$.

We now proceed to the construction of a measure in K. The expansion (1) establishes a one-to-one correspondence between the elements of K and the space E of the infinite sequences

$$\varepsilon = (\varepsilon_l, \varepsilon_{l+1}, \dots), \; \varepsilon_s \in T \qquad (s = l, l+1, \dots). \tag{2}$$

We shall define a measure on E, and this, by virtue of the one-to-one correspondence between K and E, will induce a measure on K.

Let $\{\epsilon_l^*, \epsilon_{l+1}^*, \cdots, \epsilon_k^*\}$ be an arbitrary finite set of elements of T. A set $M = M(\epsilon_l^*, \epsilon_{l+1}^*, \cdots, \epsilon_k^*)$ of all sequences (2) satisfying conditions of the form $\epsilon_l = \epsilon_l^*, \epsilon_{l+1} = \epsilon_{l+1}^*, \cdots, \epsilon_k = \epsilon_k^*$ shall be called an elementary cylinder in E. In other words, M is the set of all sequences (2) which satisfy the conditions

$$\begin{cases} \varepsilon_s = 0 & \text{for } s < l, \\ \varepsilon_s = \varepsilon_s^* & \text{for } l \leqslant s \leqslant k, \\ \varepsilon_s & \text{arbitrary for } s > k. \end{cases}$$

We let

$$\omega_0 = \varepsilon_l^* \pi^l + \varepsilon_{l+1}^* \pi^{l+1} + \dots + \varepsilon_k^* \pi^k. \tag{3}$$

It follows from (1) that the set M coincides with the circle $\{\omega \mid |\omega - \omega_0| \leq q^{-k-1}\} = C(\omega_0, q^{-k-1})$. The point ω_0 which is uniquely determined by the equation (3) shall be called the basis of the cylinder M.

We observe that every circle $C(\omega_1, q^{-k-1})$ is identical with an elementary cylinder whose basis is defined by the conditions $\omega_1 \equiv \omega_0 \pmod{P^{k+1}}$, where ω_0 has the form (3).

By a cylinder we mean a finite union of elementary cylinders. Let I be a finite set of positive integers and, for each $i \in I$, let M_i be an elementary cylinder; furthermore let

$$M_0 = \bigcup_{i \in I} M_i. \tag{4}$$

Then the set M_0 may be expressed in the form

$$M_0 = \sum_{i \in I'} M_i, \tag{5}$$

where I' is a subset of I, i.e. M_0 is a finite union of mutually disjoint elementary cylinders. This assertion follows from Lemma 3 since, if, say, $M_i \cap M_j$ is nonempty, then either $M_i \subseteq M_j$ or $M_j \subseteq M_i$.

On the set of elementary cylinders $M = M(\epsilon_l^*, \epsilon_{l+1}^*, \cdots, \epsilon_k^*)$ we introduce a measure μM by the equation

$$\mu M = q^{-k}, \tag{6}$$

and if M_0 is the cylinder introduced by (4) and (5), we let

$$\mu M_0 = \sum_{i \in I'} \mu M_i. \tag{7}$$

It is easy to see that the set of cylinders C is closed under the formation of finite unions and finite intersections. For a finite union of cylinders this follows from (4) and (5). If N is a finite intersection of cylinders

$$N = \bigcap_{l \in L} N_l, \quad N_l = \bigcup_{i \in I_l} M_i,$$

where L is a finite set of positive integers and each M_i is an elementary cylinder, then we obtain from Lemma 3 the relation

$$N = \bigcap_{l \in L} \sum_{i \in I_l'} M_i = \sum_{i \in J} M_i.$$

Here $J \subseteq \bigcup_{l \in L} I_l'$, i.e. J is a finite set and therefore N is a cylinder. Thus the set C of all cylinders forms a field and the nonnegative, additive set functions defined by the equations (6) and (7) can be shown to be σ-additive on C; hence it is a measure on this field.

By a well-known theorem of Carathéodory (see, for example, [39], Chapter I, §4.1) a measure defined on a field C has a unique extension to the smallest σ-field B containing C. For $M \in B$ one defines

$$\mu\, \mathbf{M} = \inf \sum_{i \in I} \mu M_i, \tag{8}$$

where I is a denumerable set of indices, $M_i \in \mathrm{C}$, $\mathbf{M} \subseteq \bigcup_{i \in I} M_i$ and the infimum is taken with respect to all denumerable coverings of the set $\mathbf{M}$ by cylinders from C. Then μ is a measure on the σ-field B and an outer measure on the set of all subsets of the space K.

Henceforth all metrical statements shall refer to the measure μ just defined.

§3. PROPERTIES OF THE MEASURE

In this section we will show that the outer measure μ is invariant with respect to translations by elements from K and that under the mapping $\omega \to 1/\omega$ sets of positive measure are mapped onto sets of positive measure.

Lemma 4. *Let E be a measurable set of points from K, $\omega_0 \in K$, $E_0 = E + \omega_0$; in other words $E_0 = \{\omega + \omega_0 \mid \omega \in E\}$. Then E_0 is measurable and $\mu E_0 = \mu E$.*

Proof. We may assume that E is an elementary cylinder. In fact, since for elementary cylinders we have $\mu E_0 = \mu E$, it follows from (4) and (5) that this is also true if E is any cylinder; hence the definition (8) implies that this assertion is also true for any measurable set E.

Now let E be an elementary cylinder $C(\omega_1, q^{-k-1})$. Then E_0 is an elementary cylinder $C(\omega^*, q^{-k-1})$, where $\omega^* \equiv \omega_1 + \omega_0 \pmod{P^{k+1}}$ and thus ω^* is of the form (3). But then (6) implies the equation $\mu C(\omega^*, q^{-k-1}) = \mu C(\omega_1, q^{-k-1})$, which completes the proof of the lemma.

Lemma 5. *Let E be a measurable set whose elements satisfy the inequality*

$$a < |\omega| < A \quad (0 < a < A). \tag{9}$$

Let E' be the set of all points of the form ω^{-1} with $\omega \in E$. Then

$$A^{-2} \mu E \leqslant \mu E' \leqslant a^{-2} \mu E. \tag{10}$$

Proof. As in the proof of the preceding lemma we may again restrict ourselves to the case where E is an elementary cylinder, say, $E = C(\omega_0, q^{-k-1})$, $\omega_0 = \epsilon_l \pi^l + \cdots + \epsilon_k \pi^k$, $\epsilon_l \neq 0$. We will show that E' is also an elementary cylinder. Indeed, if $\omega \in C(\omega_0, q^{-k-1})$ then $\omega = \omega_0 + \omega_1$, $|\omega_1| \leq q^{-k-1}$, and thus

$$\left| \frac{1}{\omega} - \frac{1}{\omega_0} \right| = \frac{|\omega_1|}{|\omega_0 + \omega_1|\, |\omega_0|} = \frac{|\omega_1|}{|\omega_0|^2},$$

since $|\omega_0 + \omega_1| = \max(|\omega_0|, |\omega_1|) = |\omega_0|$. Consequently, the inequality $|\omega_1| < q^{-k-1}$ is equivalent to the inequality $|\omega^{-1} - \omega_0^{-1}| \leq q^{-k-1}|\omega_0|^{-2}$. Therefore, if

$$\omega_0^{-1} = \delta_{-l}\pi^{-l} + \ldots + \delta_{k-2l}\pi^{k-2l} + \ldots = \omega_0' + \theta,\ |\theta| \leqslant q^{-k-1+2l},$$

then $E' = C(\omega_0', q^{-k-1+2l})$. Then it follows from (6) that

$$\mu E' = q^{-k+2l} = \mu E\,|\omega_0|^{-2}. \tag{11}$$

For $\omega \in E$ we have $\omega = \omega_0 + \omega_1$, $|\omega_1| < |\omega_0|$ and therefore $|\omega| = |\omega_0|$. Furthermore, if $a < |\omega| < A$ then also $a < |\omega_0| < A$. Thus (11) implies (10).

Finally, the inequality (10) is preserved if we make the transition to a finite union of mutually disjoint elementary cylinders. Therefore, it remains valid for measurable sets E and E'.

In particular, it follows from (10) that any set of positive measure is mapped onto a set of positive measure under the mapping $\omega \to \omega^{-1}$. Indeed, if $\mu E > 0$, then there exists some circle $C(0, A)$ such that the set $E_1 = E \cap C(0, A)$ satisfies $\mu E_1 > 0$. If now E' and E_1' are the images of the sets E and E_1, respectively, under this mapping, then $E' \supseteq E_1'$. Hence it follows from (10) that $\mu E' \geq \mu E_1' \geq A^{-2}\mu E_1 > 0$.

§4. DENSITY AND MEASURE

Let M be a measurable set of points from K. Then we define the "density" $d(\mathrm{M})$ of the set M by the equation

$$d(\mathrm{M}) = \sup_{(M)} \frac{\mu(\mathrm{M} \cap M)}{\mu M}, \tag{12}$$

where the supremum is taken over the set of all elementary cylinders M.

In the case of point sets on the real line the notion of metrical density was introduced by Knopp [28], who proved an analog of the following lemma.

Lemma 6. *If* $\mu\mathrm{M} > 0$ *then* $d(\mathrm{M}) = 1$.

Proof. Since the set M is measurable it follows from (8) that for any $\epsilon > 0$ there exists a covering of the set M by cylinders $N_i (i \in I)$, i.e. $\mathrm{M} \subseteq \mathrm{U}_{i \in I} N_i$, such that

$$\mu\mathrm{M} > \sum_{i \in I} \mu N_i - \varepsilon. \tag{13}$$

By Lemma 3 the union $\mathrm{U}_{i \in I} N_i$ of cylinders may be expressed in the form

$$\bigcup_{i\in I} N_i = \sum_{j\in J} M_j,$$

where the M_j are mutually disjoint elementary cylinders. Since μ, being a measure, is sub-σ-additive, we have

$$\mu(\bigcup_{i\in I} N_i) \leqslant \sum_{i\in I} \mu N_i$$

and hence

$$\mu \mathrm{M} + \varepsilon > \mu (\bigcup_{i\in I} N_i) = \sum_{j\in J} \mu M_j. \tag{14}$$

Thus we obtain

$$\mu \mathrm{M} = \mu (\mathrm{M} \cap \sum_{j\in J} M_j) = \mu \{ \sum_{j\in J} \mathrm{M} \cap M_j \}$$

$$= \sum_{j\in J} \mu (\mathrm{M} \cap M_j).$$

It follows immediately from the definition of the density $d = d(\mathrm{M})$ that $d \cdot \mu M_j \geq \mu(\mathrm{M} \cap M_j)$. Consequently, (14) implies the inequality

$$\mu \mathrm{M} \leqslant d \sum_{j\in J} \mu M_j < d(\mu \mathrm{M} + \varepsilon).$$

Since the number $\epsilon > 0$ was arbitrary, we have $\mu\mathrm{M} \leq d\mu\mathrm{M}$. Furthermore, if $\mu\mathrm{M} > 0$ then $d \geq 1$. Finally it follows from (12) that always $0 \leq d \leq 1$; hence we have $d = 1$.

Let Ω be a circle with center at ω_0 and with an arbitrary (but fixed) radius $R = q^r$, where r is a nonnegative integer, i.e. $\Omega = C(\omega_0, q^r)$.

We shall say that the set M_0 is dense in Ω if for any $\epsilon > 0$ and for any $\omega \in \Omega$ there exists a point $m_0 \in \mathrm{M}_0$ such that $|\omega - m_0| < \epsilon$.

Lemma 7. *If the set* M *is measurable,* $d(\mathrm{M}) = 1$, *and if the set* M_0 *is dense in* Ω, *then*

$$\mu \{ (\mathrm{M} + \mathrm{M}_0) \cap \Omega \} = \mu\Omega, \tag{15}$$

i.e. almost all points $\omega \in \Omega$ *can be expressed in the form*

$$\omega = m + m_0, \ m \in \mathrm{M}, \ m_0 \in \mathrm{M}_0$$

(*here and below,* $\mathrm{M} + \mathrm{M}_0$ *denotes the Šnirel'man sum*).

Proof. Let $0 < \epsilon < 1$. It follows from the equation $d(\mathrm{M}) = 1$ that there exists an elementary cylinder O such that

$$\mu(\mathbf{M}\cap O)\geqslant\mu O(1-\varepsilon),\quad O\subseteq\Omega. \tag{16}$$

Now let $O = C(\omega_0, q^{-k})$, $k \geq 0$. Clearly, Ω may be expressed in the form

$$\Omega=\sum_{\omega_i}O(\omega_i), \tag{17}$$

where $O(\omega_i)$ is the elementary cylinder with basis ω_i of the form

$$\omega_i=\varepsilon_{-r}\pi^{-r}+\varepsilon_{-r+1}\pi^{-r+1}+\ldots+\varepsilon_{k-1}\pi^{k-1},\quad \varepsilon_j\in T.$$

The number of such points ω_i is q^{k+r}. Since the set $\mathbf{M}_0$ is dense in Ω we may choose points $m_i \in \mathbf{M}_0$ belonging to the cylinders $O(\omega_i)$, respectively. Clearly, we have $O(\omega_i) = C(m_i, q^{-k})$ and hence, by (17),

$$\Omega=\sum_i C(m_i,\ q^{-k}). \tag{18}$$

The points m_i are different from each other and the number of them is q^{k+r}.

We now consider the set

$$E(m_i)=C(\omega_0+m_i,\ q^{-k})\cap(\mathbf{M}+m_i)=\{\omega:|\omega-(\omega_0+m_i)|\leqslant q^{-k},\ \omega\in\mathbf{M}+m_i\},$$

which is obtained from the set

$$E=C(\omega_0,\ q^{-k})\cap\mathbf{M}$$

by adding m_i to all its elements. It follows from Lemma 4 that the measure μ is invariant under shifts; thus $\mu E(m_i) = ME$ for any m_i. Furthermore, by the condition (16),

$$\mu(E)=\mu(O\cap\mathbf{M})\geqslant(1-\varepsilon)\mu O,$$

and hence

$$\mu E(m_i)\geqslant(1-\varepsilon)\mu O. \tag{19}$$

Instead of the decomposition (18) we now have

$$\Omega=\sum_i C(m_i+\omega_0,\ q^{-k}). \tag{20}$$

The validity of this assertion follows from the fact that the residues $m_i + \omega_0 \pmod{P^k}$ form a full system of residues modulo P^k, since this is true for the residues m_i.

Let $M = \{m_i\}$. Then (20) implies

$$\Omega\cap(\mathbf{M}+M)=\sum_{m_i\in M}C(m_i+\omega_0,\ q^{-k})\cap(\mathbf{M}+M)\supseteq\sum_{m_i\in M}C(m_i+\omega_0,\ q^{-k})\cap(\mathbf{M}+m_i)=\sum_{m_i\in M}E(m_i).$$

Consequently, it follows from (19) that

$$\mu\{\Omega\cap(\mathbf{M}+M)\}\geqslant\sum_{m_i\in M}\mu E(m_i)\geqslant(1-\varepsilon)\mu O\sum_{m\in M}1=(1-\varepsilon)\mu O q^{r+k},$$

since there are exactly q^{r+k} different points m_i. In view of the fact that $\mu O = q^{-k+1}$ and $\mu\Omega = q^{r+1}$ we obtain

$$\mu\{\Omega\cap(\mathbf{M}+M)\}\geqslant(1-\varepsilon)\mu\Omega.$$

Since this is true for arbitrary $\epsilon > 0$, we have

$$\mu\{\Omega\cap(\mathbf{M}+\mathbf{M}_0)\}\geqslant\mu\Omega,$$

which implies the assertion (15).

§5. A LEMMA ON PARTIAL COVERINGS

Lemma 8. *Let $\epsilon > 0$ and let Δ be a measurable set of points from the field K, $\mu\Delta < \epsilon$; furthermore, let $\Lambda = \bigcup_{i=1}^{\infty}\lambda_i$ be a system of circles λ_i satisfying the conditions*

$$\mu(\lambda_i\cap\Delta)\geqslant\frac{1}{2}\mu\lambda_i \quad (i = 1, 2,\dots). \tag{21}$$

Then $\mu\Lambda < 2\epsilon$.

Proof. Lemma 3 implies the existence of a subsystem $\{\lambda_j^*\}$, $j \in J$, of the system $\{\lambda_j\}$, of mutually disjoint circles λ_j such that

$$\Lambda=\bigcup_{i=1}^{\infty}\lambda_i=\sum_{j\in J}\lambda_j^*.$$

By applying the condition (21) we find

$$\mu\Lambda=\sum_{j\in J}\mu\lambda_j^*\leqslant 2\sum_{j\in J}\mu(\lambda_j^*\cap\Delta)=2\mu\Big(\sum_{j\in J}\lambda_j^*\cap\Delta\Big)$$

$$=2\mu\Big(\Delta\cap\sum_{j\in J}\lambda_j^*\Big),$$

but obviously

$$\Delta\cap\sum_{j\in J}\lambda_j^*\subseteq\Delta,$$

and hence $\mu\Lambda < 2\mu\Delta < 2\epsilon$.

§6. A REMARK ON EXTENDING A VALUATION

It is well known that if a valuation is defined on a field K, it may be extended to a valuation on a given finite extension of K (see, for example, [78], Chapter X, §76).

Let $\{\alpha_n\}$ be a sequence of elements which are algebraic over K and let K^* be the smallest extension field of K which contains all elements of this sequence. The field K^* may be defined by the following procedure: We consider the sequence of the fields

$$K_0 = K,\ K_1 = K_0(\alpha_1), \ldots,\ K_n = K_{n-1}(\alpha_n), \ldots .$$

Trivially,

$$K_0 \subseteq K_1 \subseteq K_2 \subseteq \ldots \subseteq K_n \subseteq \ldots, \tag{22}$$

and the field K^* may be defined by the equation

$$K^* = \bigcup_{i=1}^{\infty} K_i = \lim_{n\to\infty} K_n$$

The given valuation on $K = K_0$ may be extended to a valuation on the field K_1, this being a finite extension field of K_0. Knowing the valuation on K_1 we may extend it to a valuation on K_2, etc. In this way we obtain on each field K_n $(n = 1, 2, \cdots)$ a valuation which coincides on the field $K \subseteq K_n$ with the initial valuation. Thus we have extended the given valuation on K to a valuation on the field K^*.

Since we assume the field K to be complete (with respect to the given valuation), the transition from K_i to $K_{i+1} = K_i(\alpha_{i+1})$ leads in each case to a unique extension of the valuation from K_i to K_{i+1}. Indeed, if the field K_i is complete with respect to a given valuation on this field then the finite extension field K_{i+1} is also complete with respect to the extension of the valuation on K_{i+1} defined, for each $\omega \in K_{i+1}$, by the equation

$$|\omega| = |\mathrm{Nm}(\omega)|^{\frac{1}{s}}, \tag{23}$$

where $\mathrm{Nm}(\omega)$ is the algebraic norm of the element ω in K_i and $s = s_i$ is the degree of K_{i+1} over K_i.

Furthermore, if Γ_i is the valuation group of the field K_i, then

$$\Gamma_0 \subseteq \Gamma_1 \subseteq \Gamma_2 \subseteq \ldots \subseteq \Gamma_n \subseteq \ldots,$$

by (22). Therefore, if the given valuation is discrete on K (in this case $\Gamma_0 = \Gamma$ is a discrete group) then all groups Γ_i are discrete, since it follows from (23) that all indices $(\Gamma_{i+1} : \Gamma_i)$ are finite and the valuation is discrete on all the fields K_i. On the field K^*, however, the valuation may be nondiscrete.

Finally, let K' be a finite extension field of K obtained by adjoining to K a finite number of terms from the sequence $\{\alpha_n\}$. Then there exists an index n such that $K' \subseteq K_n$. Since the valuation on K_n is uniquely determined, it is also uniquely determined as a valuation on the field K'.

In considering finite extension fields of the ground field K we shall henceforth assume that a valuation has been defined on them by the method outlined above.

§7. ESTIMATES FOR THE DISTANCE $|\omega - \kappa|$

Let $F(x) = a_0 + a_1 x + \cdots + a_n x^n$, $a_n \neq 0$, be a polynomial over K', where K' is a finite extension field of K. We assume that all the roots $\kappa_1, \kappa_2, \cdots, \kappa_n$ of the polynomial F are different from each other and numbered such that

$$|\kappa_1 - \kappa_2| \leqslant |\kappa_1 - \kappa_3| \leqslant \cdots \leqslant |\kappa_1 - \kappa_n|. \tag{24}$$

For each root κ_i we define the set $S(\kappa_i)$ of those points ω in the field K which satisfy the condition

$$\min_{(j)} |\omega - \kappa_j| = |\omega - \kappa_i| \quad (j = 1, 2, \ldots, n). \tag{25}$$

Lemma 9. *If* $\omega \in S(\kappa_1)$ *then*

$$|\omega - \kappa_1| \leqslant \min\left(\frac{|F(\omega)|}{|F'(\kappa_1)|}, \left(\frac{|F(\omega)|}{|F'(\kappa_1)|}|\kappa_1 - \kappa_2|\right)^{\frac{1}{2}}\right). \tag{26}$$

Proof. It follows from (25) that

$$|\kappa_1 - \kappa_i| \leqslant \max(|\kappa_1 - \omega|, |\kappa_i - \omega|) = |\omega - \kappa_i| \; (i = 2, 3. \ldots, n).$$

Therefore,

$$|F'(\kappa_1)| = |a_n| \prod_{i=2}^{n} |\kappa_1 - \kappa_i| \leqslant |a_n| \prod_{i=2}^{n} |\omega - \kappa_i| = \frac{|F(\omega)|}{|\omega - \kappa_1|},$$

and hence

$$|\omega - \kappa_1| \leqslant \frac{|F(\omega)|}{|F'(\kappa_1)|} \tag{27}$$

Similarly, we obtain from (25)

$$\frac{|F'(\kappa_1)|}{|\kappa_1 - \kappa_2|} = |a_n| \prod_{i=3}^{n} |\kappa_1 - \kappa_i| \leqslant |a_n| \prod_{i=3}^{n} |\omega - \kappa_i|$$

$$= \frac{|F(\omega)|}{|\omega - \kappa_1||\omega - \kappa_2|},$$

and since $|\omega - \kappa_1| \leq |\omega - \kappa_2|$, we have

$$|\omega - \varkappa_1|^2 \leqslant |\omega - \varkappa_1||\omega - \varkappa_2| \leqslant \frac{|F(\omega)|}{|F'(\varkappa_1)|}|\varkappa_1 - \varkappa_2|,$$

$$|\omega - \varkappa_1| \leqslant \left(\frac{|F(\omega)|}{|F'(\varkappa_1)|}|\varkappa_1 - \varkappa_2|\right)^{\frac{1}{2}}. \tag{28}$$

The assertion (26) is an immediate consequence of (27) and (28).

Lemma 10. *If* $\omega \in S(\kappa_1)$ *and* $|\omega - \kappa_1| \leq |\kappa_1 - \kappa_2|$ *then*

$$|\omega - \varkappa_1| = \frac{|F(\omega)|}{|F'(\varkappa_1)|}. \tag{29}$$

If $|\omega - \kappa_1| \geq |\kappa_1 - \kappa_2|$ *and if furthermore*

$$|\omega - \varkappa_1| \leqslant |\varkappa_1 - \varkappa_3|, \tag{30}$$

then

$$|\omega - \varkappa_1| = \left(\frac{|F(\omega)|}{|F'(\varkappa_1)|}|\varkappa_1 - \varkappa_2|\right)^{\frac{1}{2}}. \tag{31}$$

Proof. If $|\omega - \kappa_1| \leq |\kappa_1 - \kappa_2|$ then we have

$$|\omega - \varkappa_i| \leqslant \max(|\omega - \varkappa_1|, |\varkappa_1 - \varkappa_i|)$$

$$\leqslant \max(|\varkappa_1 - \varkappa_2|, |\varkappa_1 - \varkappa_i|) = |\varkappa_1 - \varkappa_i| \quad (i = 2, 3, \ldots, n).$$

Hence

$$\frac{|F(\omega)|}{|\omega - \varkappa_1|} = |a_n| \prod_{i=2}^{n} |\omega - \varkappa_i| \leqslant |a_n| \prod_{i=2}^{n} |\varkappa_1 - \varkappa_i| = |F'(\varkappa_1)|$$

and consequently,

$$|\omega - \varkappa_1| \geqslant \frac{|F(\omega)|}{|F'(\varkappa_1)|}. \tag{32}$$

In this case we obtain the assertion (29) by applying (27).

Now suppose that $|\omega - \kappa_1| \geq |\kappa_1 - \kappa_2|$ and that (30) is satisfied. Then it follows from (30) and (24) that

$$|\omega - \varkappa_i| \leqslant \max(|\omega - \varkappa_1|, |\varkappa_1 - \varkappa_i|) \leqslant |\varkappa_1 - \varkappa_i| \; (i = 3, 4, \ldots, n).$$

Thus

$$\prod_{i=3}^{n} |\omega - \varkappa_i| \leqslant \prod_{i=3}^{n} |\varkappa_1 - \varkappa_i|,$$

and since

$$|\omega - \varkappa_2| \leqslant \max(|\omega - \varkappa_1|, |\varkappa_1 - \varkappa_2|) = |\omega - \varkappa_1|.$$

we obtain

$$|F(\omega)| \leqslant |\omega - \varkappa_1||\omega - \varkappa_2| \frac{|F'(\varkappa_1)|}{|\varkappa_1 - \varkappa_2|} \leqslant |\omega - \varkappa_1|^2 \frac{|F'(\varkappa_1)|}{|\varkappa_1 - \varkappa_2|}.$$

So we obtain in this case the inequality

$$|\omega - \varkappa_1| \geqslant \left(\frac{|F(\omega)|}{|F'(\varkappa_1)|} |\varkappa_1 - \varkappa_2|\right)^{\frac{1}{2}}. \tag{33}$$

Together with (28) this yields the assertion (31).

§8. THE STRUCTURE OF THE DOMAINS $\sigma_i(F)$

Let $\tau > 0$ be a real number. We denote by $\sigma(F) = \sigma(F, \tau)$ the set of those points ω of the field K for which $|F(\omega)| \leq \tau$, where F is the polynomial defined at the beginning of the preceding section. Furthermore we define

$$\sigma_i(F) = \sigma(F) \cap S(\varkappa_i) \quad (i = 1, 2, \ldots, n).$$

Since it is trivial to see how the properties of the domains $\sigma_i(F)$ depend on the index i we shall consider only one such domain, say $\sigma_1(F)$. In the sequel we shall assume that $\sigma_1(F)$ is nonempty and that all its points satisfy the condition (30).

Lemma 11. *Let ω_0 be a point of K which satisfies the condition*

$$|\omega_0 - \varkappa_1| = \max |\omega - \varkappa_1|,$$

where the maximum is taken over all points ω from the domain $\sigma_1(F)$. Then the domain $\sigma_1(F)$ is equal to the circle

$$|\omega - \varkappa_1| \leqslant |\omega_0 - \varkappa_1| = \frac{|F(\omega_0)|}{|F'(\varkappa_1)|} \tag{34}$$

if (*) $|\omega_0 - \kappa_1| < |\kappa_1 - \kappa_2|$; *in the case* (**) $|\omega_0 - \kappa_1| \geq |\kappa_1 - \kappa_2|$ *the domain $\sigma_1(F)$ is the intersection of $S(\kappa_1)$ with the circle*

$$|\omega - \varkappa_1| \leqslant |\omega_0 - \varkappa_1| = \left(\frac{|F(\omega_0)|}{|F'(\varkappa_1)|} |\varkappa_1 - \varkappa_2|\right)^{\frac{1}{2}}. \tag{35}$$

Proof. First we observe that the function $f(\omega) = |\omega - \kappa_1|$ assumes a maximum in the domain $\sigma_1(F)$; in other words, there exists a point ω_0 with the property mentioned in the lemma. This fact follows immediately from basic theorems on continuous functions over a non-archimedean field (see, for example, [35], pp. 54–55). Indeed, $F(x)$ is a continuous function of $x \in K$ and therefore $\sigma(F)$ is closed in the natural topology of the field K. Also it is evident that the sets

$$S_i = \{\omega : |\omega - \varkappa_1| \leqslant |\omega - \varkappa_i|\} \quad (i = 2, 3, \ldots, n) \tag{36}$$

and the set

$$S(\varkappa_1) = S_2 \cap S_3 \cap \ldots \cap S_n \tag{37}$$

are closed. Consequently, the set $\sigma_1(F) = \sigma(F) \cap S(\kappa_1)$ is closed. Finally, the set $\sigma_1(F)$ is bounded since it follows from the inequality $|F(\omega)| \leq \tau$ that for all $\omega \in S(\kappa_1)$ one has

$$|\omega - \varkappa_1|^n |a_n| \leqslant |F(\omega)| \leqslant \tau,$$

$$|\omega - \varkappa_1| \leqslant \left(\frac{\tau}{|a_n|}\right)^{\frac{1}{n}}.$$

Therefore the continuous function $f(\omega) = |\omega - \kappa_1|$ does indeed assume a maximal value on the closed set $\sigma_1(F)$.

In the case (*) we show that the condition $\omega \in K$, $|\omega - \kappa_1| \leq |\omega_0 - \kappa_1|$ implies $\omega \in S(\kappa_1)$, $|F(\omega)| \leq |F(\omega_0)|$; in other words, the circle (34) is contained in the set $\sigma_1(F)$.

Suppose $|\omega - \kappa_j| < |\omega - \kappa_1|$ for such a point ω. Then

$$|\varkappa_1 - \varkappa_j| = |\varkappa_1 - \omega - (\varkappa_j - \omega)| = |\varkappa_1 - \omega| \leqslant |\omega_0 - \varkappa_1| < |\varkappa_1 - \varkappa_2|,$$

but this is impossible because of the convention (24). Therefore $|\omega - \kappa_j| \geq |\omega - \kappa_1|$ ($j = 2, 3, \cdots, n$) and hence $\omega \in S(\kappa_1)$. Because of the inequality $|\omega_0 - \kappa_1| < |\kappa_1 - \kappa_2|$, in view of the assertion (29) from Lemma 10, this implies that

$$|\omega - \varkappa_1| = \frac{|F(\omega)|}{|F'(\varkappa_1)|}, \quad |\omega_0 - \varkappa_1| = \frac{|F(\omega_0)|}{|F'(\varkappa_1)|},$$

and therefore one has $|F(\omega)| \leq |F(\omega_0)|$. But since $\omega_0 \in \sigma_1(F)$, it follows that $|F(\omega_0)| \leq \tau$ and hence $\omega \in \sigma(F)$. So we may conclude that the entire circle (34) is indeed contained in the domain $\sigma_1(F)$.

The converse assertion is obvious, since by the choice of the point ω_0 all points $\omega \in \sigma_1(F)$ satisfy the inequality $|\omega - \kappa_1| \leq |\omega_0 - \kappa_1|$.

In the case (**) it suffices in view of the last observation to show that the domain $\sigma_1(F)$ contains the intersection of the set $S(\kappa_1)$ and the circle (35); in other words, that the conditions

$$\omega \in S(\varkappa_1), \ |\omega - \varkappa_1| \leqslant |\omega_0 - \varkappa_1| \tag{38}$$

imply $\omega \in \sigma_1(F)$.

First we assume that $|\omega - \kappa_1| \geq |\kappa_1 - \kappa_2|$. Then it follows from (31) by

Lemma 10 that

$$|\omega - \kappa_1| = \left(\frac{|F(\omega)|}{|F'(\kappa_1)|} |\kappa_1 - \kappa_2| \right)^{\frac{1}{2}}.$$

The analogous equation is true for ω_0, i.e.

$$|\omega_0 - \kappa_1| = \left(\frac{|F(\omega_0)|}{|F'(\kappa_1)|} |\kappa_1 - \kappa_2| \right)^{\frac{1}{2}}.$$

Thus we obtain $|F(\omega)| \leq |F(\omega_0)|$ by (38), and consequently $\omega \in \sigma(F)$. Since we are assuming that $\omega \in S(\kappa_1)$, this implies $\omega \in \sigma_1(F)$.

Now let $|\omega - \kappa_1| < |\kappa_1 - \kappa_2|$ and at the same time $|\omega_0 - \kappa_1| \geq |\kappa_1 - \kappa_2|$. Since $\omega \in S(\kappa_1)$ we then obtain from (29) the relation

$$|\omega - \kappa_1| = \frac{|F(\omega)|}{|F'(\kappa_1)|} < |\kappa_1 - \kappa_2|,$$

$$|F(\omega)| < |F'(\kappa_1)| \, |\kappa_1 - \kappa_2|. \tag{39}$$

On the other hand, the quantity $|\omega_0 - \kappa_1|$ satisfies, because of (31), the relation

$$|\kappa_1 - \kappa_2| \leqslant |\omega_0 - \kappa_1| = \left(\frac{|F(\omega_0)|}{|F'(\kappa_1)|} |\kappa_1 - \kappa_2| \right)^{\frac{1}{2}},$$

and thus we have

$$|F(\omega_0)| \geqslant |F'(\kappa_1)| \, |\kappa_1 - \kappa_2|. \tag{40}$$

From (39) and (40) we may conclude that all ω from the set $\sigma(F)$, and hence all ω from $\sigma_1(F)$, satisfy the inequality $|F(\omega)| < |F(\omega_0)|$.

Thus it is clear that the intersection of $S(\kappa_1)$ and the circle (35) is contained in $\sigma_1(F)$.

Lemma 12. *The set $\sigma(F)$ and also the sets $\sigma_i(F)$ $(i = 1, 2, \cdots, n)$ are measurable (with respect to the measure introduced in §2).*

Proof. First we observe that, if κ is algebraic over K and if r is a positive real number, then the set C of points $\omega \in K$ satisfying $|\omega - \kappa| \leq r$ is a circle.

Indeed, let us define $\overset{0}{\kappa}$ by the condition $|\overset{0}{\kappa} - \kappa| = \min |\omega - \kappa|$, the minimum being taken over all $\omega \in K$. Clearly, if $\kappa \in K$, then $\overset{0}{\kappa} = \kappa$, but it is certainly possible that $\kappa \notin K$. Let us assume that C is a circle with center at $\overset{0}{\kappa}$ and with radius r_0 equal to the largest number of the form q^k not exceeding r, i.e. $C = C(\overset{0}{\kappa}, q^k)$. For any $\omega \in K$ we have $|\omega - \kappa| \geq |\overset{0}{\kappa} - \kappa|$ and hence

$$|\omega - \overset{0}{\varkappa}| = |\omega - \varkappa - (\overset{0}{\varkappa} - \varkappa)| \leqslant \max(|\omega - \varkappa|, |\overset{0}{\varkappa} - \varkappa|) = |\omega - \varkappa|.$$

Consequently, if $|\omega - \kappa| \leq r$, then $|\omega - \overset{0}{\kappa}| \leq r$, i.e. $C \subseteq C(\overset{0}{\kappa}, q^k)$. Conversely, if $|\omega - \overset{0}{\kappa}| \leq r$, then

$$|\omega - \varkappa| = |\omega - \overset{0}{\varkappa} - (\varkappa - \overset{0}{\varkappa})| \leqslant \max(|\omega - \overset{0}{\varkappa}|, |\overset{0}{\varkappa} - \varkappa|).$$

By assumption, the set C is nonempty, and it follows from the construction of $\overset{0}{\kappa}$ that we have $|\overset{0}{\kappa} - \kappa| \leq r$. Therefore,

$$|\omega - \varkappa| \leqslant \max(|\omega - \overset{0}{\varkappa}|, |\overset{0}{\varkappa} - \varkappa|) \leqslant r.$$

In the case where $|\omega - \overset{0}{\kappa}| \leq r$, we have therefore $|\omega - \kappa| \leq r$, i.e. $C(\overset{0}{\kappa}, q^k) \subseteq C$, which implies the equality $C = C(\overset{0}{\kappa}, q^k)$.

Secondly, let K_F be the splitting field of the polynomial F over K' and let $(K_F : K') = m$. Because of (23) the norm of an element $\alpha \in K_F$ has the form $\{\alpha\} = q^{k/m}$, where k is an integer. In particular, this is true for $\alpha = \omega - \kappa_i$ $(i = 1, 2, \cdots, n)$, where the numbers κ_i are the roots of F.

For any integer k we consider the sets

$$M_k(\varkappa_i) = \{\omega \in K : |\omega - \varkappa_i| \leqslant q^{\frac{k}{m}}\} \quad (i = 1, 2, \ldots, n), \tag{41}$$

$$N_k(\varkappa_i) = \{\omega \in K : |\omega - \varkappa_i| \geqslant q^{\frac{k}{m}}\} \quad (i = 1, 2, \ldots, n). \tag{42}$$

It follows from the remarks made above that the set $M_k(\kappa_i)$ coincides with some circle in K; hence it is measurable (any circle in K is identical to an elementary cylinder whose measure is defined by the equation (6)). Furthermore, if $r \in \Gamma$ and if $C(0, r)$ is a circle in K, then

$$N_k(\varkappa_i) \cap C(0, r) = C(0, r) \setminus M_{k-1}(\varkappa_i), \tag{43}$$

since the set $N_k(\kappa_i) \cap C(0, r)$ consists of the points $\omega \in K$ satisfying $r \geq |\omega - \kappa_i| \geq q^{k/m}$, i.e. of all points $\omega \in K$ of the circle $C(0, r)$ except those for which $|\omega - \kappa_i| < q^{k/m}$. Thus the set $N_k(\kappa_i)$ is also measurable by (43).

Finally, since the inequality $|F(\omega)| \leq \tau$ is equivalent to

$$\prod_{i=1}^{n} |\omega - \varkappa_i| \leqslant \frac{\tau}{|a_n|}, \tag{44}$$

the set $\sigma(F)$ may be expressed in the form

$$\sigma(F) = \bigcup_{\substack{(k) \\ k_1+k_2+\dots+k_n=mk}} \bigcap_{i=1}^{n} M_{k_i}(\varkappa_i) \cap N_{k_i}(\varkappa_i), \tag{45}$$

where the union is taken over all systems $(k_1, k_2, \cdots, k_n)$ of positive integers which satisfy the condition $k_1 + k_2 + \cdots + k_n = mk$, and where k runs through all positive integers with $q^{k/m} \leq r|a_n|^{-1}$.

Since the measurable sets form a σ-field (also called a σ-algebra) it follows from (45) and the measurability of the sets (41) and (42) as shown above that the set $\sigma(F)$ is also measurable.

The measurability of the sets $\sigma_i(F)$ follows from the fact that the sets $S(\kappa_i)$ $(i = 1, \cdots, n)$ are measurable. Indeed, we have $\sigma_i(F) = \sigma(F) \cap S(\kappa_i)$ $(i = 1, \cdots, n)$, and the measurability of the sets $S(\kappa_i)$ is established as follows: Let the sets

$$T_{ij} = \{\omega : |\omega - \varkappa_i| \leqslant |\omega - \varkappa_j|\} \quad (i, j = 1, 2, \dots, n),$$

be such that

$$S(\varkappa_i) = \bigcap_{\substack{1 \leqslant j \leqslant n \\ j \neq i}} T_{ij}.$$

In view of the last equation it suffices to establish the measurability of the sets T_{ij}. By (41) and (42) we have

$$T_{ij} = \bigcup_{-\infty < k < \infty} M_k(\varkappa_i) \cap N_k(\varkappa_j),$$

where the union is taken over all integers k. But since the sets $M_k(\kappa_i)$ and $N_k(\kappa_i)$ are measurable, this is also true for the set T_{ij}. Thus the proof of the lemma is completed.

§9. CONCLUSION

It is well known that there are, essentially, only two types of locally compact fields K with non-archimedean valuation: (1) finite extensions of Hensel's fields of p-adic numbers and (2) fields over formal power series over finite fields (see, for example, [35], pp. 44–46). Thus, inasmuch as our basic problem is by its nature a metrical one and inasmuch as the existence of a measure on K induced by the natural topology on the field can be guaranteed only if the field K is locally compact, it is evident that it suffices in our further investigation to restrict ourselves to fields of the two indicated types (1) and (2). The following two chapters are devoted to the cases (1) and (2), respectively. In connection with case (1) we consider immediately the field Q_p of p-adic numbers, since the analogous problem for finite extensions of Q_p does then not require any additional essential ideas. In both chapters the arguments will be based on the present chapter.

CHAPTER 2

FIELDS OF p-ADIC NUMBERS

§1. DIOPHANTINE APPROXIMATION IN Q_p

In this chapter we are considering an analog of Mahler's conjecture for p-adic numbers, i.e. for the elements of the field Q_p. In Q_p there exists a p-adic valuation which will be denoted by the traditional notation $|\omega|_p$ for all $\omega \in Q_p$. The definition and basic properties of p-adic numbers may be found, for example, in the book [79] by H. Weyl or in Borevič-Šafarevič [3].

In the sequel we shall need some simple results from the p-adic theory of diophantine approximations. The main theorems in this direction were established by Mahler [43] (see also the monograph [40] by E. Lutz). In particular, the following lemma is implied by some results contained in the paper [43].

Lemma 1. *Let* $n+1$ *linear forms*

$$L_i(x_0, x_1, \ldots, x_n) = L_i(\mathbf{x}) \qquad (i = 0, 1, \ldots, n)$$

with real coefficients and with determinant $d \neq 0$ *be given. Furthermore, let* $L(\mathbf{x})$ *be a linear form with integer coefficients. Finally, let* $\beta, \beta_0, \cdots, \beta_n$ *be positive real numbers with* $0 < \beta < 1$ *and* $\beta\beta_0\beta_1 \cdots \beta_n \geq p|d|$. *Then there exists a vector* $\mathbf{x} \neq (0)$ *with (rational) integer components such that the system of inequalities*

$$|L(\mathbf{x})|_p < \beta,$$
$$|L_i(\mathbf{x})| \leqslant \beta_i \qquad (i = 0, 1, \ldots, n).$$

is satisfied.

As an immediate consequence of this lemma we obtain the following one:

Lemma 2. *Let* $H > 1$ *be an arbitrary integer and* $L(\mathbf{x})$ *a linear form whose coefficients are p-adic integers. Then there exists a vector* $\mathbf{x} = (x_0, \cdots, x_n) \neq (0)$, *with integer components,*

$$\max_{(i)} |x_i| \leq H \qquad (i = 0, 1, \ldots, n),$$

such that the inequality

$$|L(\mathbf{x})|_p < pH^{-n-1} \tag{46}$$

is satisfied.

Proof. In the preceding lemma let $L_i(\mathbf{x}) = x_i$, $\beta_i = H$ $(i = 0, 1, \cdots, n)$ and $\beta = pH^{-n-1}$. The assertion may be obtained by a straightforward application of Dirichlet's pigeonhole principle. Indeed, if δ is the integer determined by the condition $p^{\delta} \leq H^{n+1} < p^{\delta+1}$ then it is obvious that there exist at least $([H]+1)^{n+1} > H^{n+1}$ integral vectors $\mathbf{x}$ which satisfy $0 \leq x_i \leq H$. Hence there exists a pair of different vectors $\mathbf{x}_1, \mathbf{x}_2$ with $L(\mathbf{x}_1) \equiv L(\mathbf{x}_2) \pmod{p^{\delta}}$. Then the vector $\mathbf{X} = \mathbf{x}_1 - \mathbf{x}_2$ satisfies the inequality (46) since

$$|L(\mathbf{X})|_p = |L(\mathbf{x}_1 - \mathbf{x}_2)|_p \leq p^{-\delta} < pH^{-n-1}.$$

Lemma 3. *For all* $\omega \in K_p$ *let* $w_n(\omega)$ *denote the supremum of the set of those* $w > 0$ *for which there exist infinitely many integral vectors* $(a_0, a_1, \cdots, a_n)$ *satisfying the inequality*

$$|a_0 + a_1\omega + \ldots + a_n\omega^n|_p < h^{-w},$$

where

$$h = \max(|a_0|, \ldots, |a_n|).$$

Then $w_n(\omega) \geq n+1$ $(n = 1, 2, \cdots)$.

Proof. If $|\omega|_p = p^k$ then $\omega_1 = \omega p^{-k}$ is a p-adic integer. Applying Lemma 2 to the linear form

$$L(\mathbf{x}) = x_0 + x_1\omega_1 + \ldots + x_n\omega_1^n$$

we obtain $w_n(\omega_1) \geq n+1$, and this obviously implies the assertion $w_n(\omega) \geq n+1$.

§2. LEMMAS ON POLYNOMIALS

In this section we denote by $\mathrm{F}_n(h)$ the set of polynomials $F(x) = a_0 + a_1x + \cdots + a_nx^n$, $n \geq 2$, with simple roots $\kappa_1, \kappa_2, \cdots, \kappa_n$ whose coefficients are rational integers subject to the conditions

$$\max(|a_0|, |a_1|, \ldots, |a_{n-1}|) \leqslant a_n = h, \ |h|_p > p^{-n}. \tag{47}$$

Furthermore, let $\overline{\mathrm{F}}_n(h)$ be the set of those polynomials from $\mathrm{F}_n(h)$ which are of the form

$$F_i(x) = F(x)(x - \varkappa_i)^{-1}, \tag{48}$$

for some index i.

Let Q_p^* be the smallest field containing Q_p and all algebraic numbers. Since the set of all algebraic numbers is denumerable, it follows from the considerations of §6 of the preceding chapter that the p-adic valuation Q_p may be extended to a valuation on Q_p^*. We shall use the notation $|\alpha|_p$ for all $\alpha \in Q_p^*$.

Lemma 4. *If $F \in \mathrm{F}_n(h)$ then*

$$\max_{(i)} |\varkappa_i|_p < p^{-n} \qquad (i = 1, 2, \ldots, n). \tag{49}$$

Proof. Evidently, $h\kappa_i$ is an algebraic integer and hence $|h\kappa_i|_p \leq 1$. Thus the assertion follows from the fact that $|h|_p > p^{-n}$.

Lemma 5. *If $F \in \overline{\mathrm{F}}_n(h)$ and if $D(F)$ denotes the discriminant of F then the inequality*

$$|F'(\varkappa)|_p > c(n, p) |D(F)|_p^{\frac{1}{2}}, \tag{50}$$

holds for any root κ of the polynomial F.

Proof. Expressing the discriminant $D(F)$ in the form

$$D(F) = h^{2n-2} \prod_{1 \leqslant i < j \leqslant n} (\varkappa_i - \varkappa_j)^2, \tag{51}$$

we obtain

$$|D(F)|_p = |h|_p^{2n-2} \prod_{1 \leqslant i < j \leqslant n} |\varkappa_i - \varkappa_j|_p^2 =$$

$$= |h|_p^2 \prod_{i \neq 1} |\varkappa_1 - \varkappa_i|_p^2 |h|_p^{2n-4} \prod_{\substack{i \neq j \\ i,j \neq 1}} |\varkappa_i - \varkappa_j|_p. \tag{52}$$

On the other hand, by Lemma 4 we have

$$|\varkappa_i - \varkappa_j|_p \leqslant \max(|\varkappa_i|_p, |\varkappa_j|_p) < p^n \quad (i, j = 1, 2, \ldots, n). \tag{53}$$

and hence

$$|D(F)|_p \leqslant c(n, p) |F'(\varkappa_1)|_p^2,$$

which implies the assertion (50).

Lemma 6. *Let* $\omega \in K_p$, $F \in \overline{\mathrm{F}}_n(h)$, *and let*

$$|\omega - \varkappa_1|_p = \min_{(i)} |\omega - \varkappa_i|_p \qquad (i = 1, 2, \ldots, n). \tag{54}$$

Then

$$|\omega - \varkappa_1|_p < c(n, p) \frac{|F(\omega)|_p}{|D(F)|_p^{1/2}}. \tag{55}$$

Proof. The assertion follows from Lemma 9 of the preceding chapter and from Lemma 5.

Lemma 7. *Under the conditions of Lemma 6 let*

$$|F(\omega)|_p < h^{-w}, \quad w > 0. \tag{56}$$

Then the inequality

$$|\omega - \varkappa_1|_p < c(n, p) h^{-\frac{2}{3}w} |D(F)|_p^{-1/6}$$

holds for any $n > 2$.

Proof. The expression (52) may be written in the form

$$|D(F)|_p = |F'(\varkappa_1)|_p^2 |D(F_1)|_p,$$

where $F_1 = F(x)(x - \kappa_1)^{-1}$. Thus the polynomial F_1 belongs to the set $\overline{\mathrm{F}}_n(h)$ and is of the form (48). Let $\rho > 0$ be a fixed real number and suppose that $|D(F_1)|_p < h^{-2\rho}$. Then

$$|D(F)|_p < |F'(\varkappa_1)|_p^2 h^{-2\rho},$$

and thus

$$|F'(\varkappa_1)|_p > h^{\rho} |D(F)|_p^{\frac{1}{2}}.$$

Consequently, by Lemma 9 of the preceding chapter and by (56), we have

$$|\omega - \varkappa_1|_p < h^{-w-\rho} |D(F)|_p^{\frac{1}{2}} \tag{57}$$

On the other hand, if $|D(F_1)|_p \geq h^{-2\rho}$, we apply Lemma 6 to the polynomial $F_1(x)$, assuming that

$$|\omega - \varkappa_2|_p = \min_{(i)} |\omega - \varkappa_i|_p \qquad (i = 2, 3, \ldots, n).$$

Then (55) yields

$$\begin{aligned} |\omega - \varkappa_2|_p &< c(n, p) |F_1(\omega)|_p |D(F_1)|_p^{-\frac{1}{2}} \\ &\leqslant c(n, p) h^{-w+\rho} |\omega - \varkappa_1|_p^{-1}. \end{aligned} \tag{58}$$

Thus we obtain instead of (57) the inequality

$$|\omega - \varkappa_1|_p < c(n, p) \max \left(h^{-w-\rho} |D(F)|_p^{-\frac{1}{2}}, h^{-\frac{1}{2}(w-\rho)}\right), \tag{59}$$

where the fact has been used that (58) implies

$$|\omega - \varkappa_1|_p^2 \leqslant |\omega - \varkappa_1|_p |\omega - \varkappa_2|_p < c(n, p) h^{-w+\rho}$$

Now we make the two quantities within the parantheses of (59) equal to each other by an appropriate choice of ρ; this leads to the assertion of the lemma.

Lemma 8. *Let the roots of the polynomial $F \in \mathrm{F}_n(h)$ be numbered so that*

$$|\varkappa_1 - \varkappa_2|_p \leqslant |\varkappa_1 - \varkappa_3|_p \leqslant \ldots \leqslant |\varkappa_1 - \varkappa_n|_p. \tag{60}$$

Furthermore, let $\omega \in K_p$ be an element satisfying the conditions (54) *and* (56) *for some $w \geq n > 2$. Then there exists a constant $c(n, p)$ such that*

$$|\omega - \varkappa_1|_p \leqslant |\varkappa_1 - \varkappa_3|_p \tag{61}$$

for all $h > c(n, p)$.

Proof. Suppose (61) is violated, i.e. $|\omega - \kappa_1|_p > |\kappa_1 - \kappa_3|_p$. Then (60) implies

$$\prod_{1 \leqslant i < j \leqslant 3} |\varkappa_i - \varkappa_j|_p < |\omega - \varkappa_1|_p^3.$$

By Lemma 7 we have, under the present conditions,

$$|\omega - \varkappa_1|_p < c(n, p) h^{-\frac{2}{3}n} |D(F)|_p^{-1/6},$$

and thus

$$\prod_{1\leqslant i<j\leqslant 3} |\varkappa_i - \varkappa_j|_p < c(n,\ p)\, h^{-2n} |D(F)|_p^{-1/2}. \tag{62}$$

On the other hand, since $F(x)$ does not have multiple roots, we have $D(F) \neq 0$. Also, clearly $|D(F)| < c(n)h_F^{2n-2}$; thus $|D(F)|_p > c(n)h_p^{-2n+2} = c(n)h^{-2n+2}$. Together with (51) and (49) this yields the inequality

$$1 < c(n)\, h^{2n-2} |D(F)|_p < c(n,\ p)\, h^{2n-2} \prod_{1\leqslant i<j\leqslant n} |\varkappa_i - \varkappa_j|_p^2$$

$$< c(n,\ p)\, h^{2n-2} \prod_{1\leqslant i<j\leqslant 3} |\varkappa_i - \varkappa_j|_p^2.$$

Consequently,

$$\prod_{1\leqslant i<j\leqslant 3} |\varkappa_i - \varkappa_j|_p > c(n,\ p)\, h^{-n+1}.$$

But since $|D(F)|_p > c(n)h^{-2n+2}$ it follows from the inequality (62) that

$$\prod_{1\leqslant i<j\leqslant 3} |\varkappa_i - \varkappa_j|_p < c(n,\ p)\, h^{-n-1}.$$

For $h > c(n,\ p)$ the last two inequalities contradict each other, and hence (61) follows.

Lemma 9. *Let* $F \in \mathbf{F}_n(h)$ *and let* m_0 *be a positive integer. Then*

$$\max_{m_0\leqslant m\leqslant m_0+n} |F(m)|_p > p^{-n} \max_{0\leqslant i\leqslant n} |a_i|_p, \tag{63}$$

where the maximum is taken over all integers $m = m_0,\ m_0 + 1, \cdots, m_0 + n$ *and over* $i = 0,\ 1, \cdots, n$.

Proof. Rewriting $F(x)$ as an interpolation polynomial with respect to the values $F(m_0),\ F(m_0 + 1), \cdots, F(m_0 + n)$, we obtain

$$F(x) = \sum_{m=m_0}^{m_0+n} \frac{A(x)}{(x-m)\, A'(m)}\, F(m),$$

with the abbreviation

$$A(x) = (x - m_0) \ldots (x - m_0 - n).$$

Then a comparison of coefficients between this expansion and the expansion

$F(x) = a_0 + a_1 x + \cdots + a_n x^n$ leads to

$$a_i = \sum_{m=m_0}^{m_0+n} \frac{B_i(m)}{A'(m)} F(m) \qquad (i = 0, 1, \ldots, n),$$

where the numbers $B_i(m)$ are integers. Consequently,

$$|a_i|_p \leqslant \max_{m_0 < m \leqslant m_0+n} \frac{|F(m)|_p}{|A'(m)|_p} \qquad (i = 0, 1, \ldots, n).$$

But since

$$A'(m) = \prod_{\substack{m_0 \leqslant l \leqslant m_0+n \\ l \neq m}} (m - l) = \pm (m - m_0)!\,(m_0 + n - m)!,$$

it follows that $A'(m)$ is divisible by power p^α, where the exponent α does not exceed the quantity

$$\left[\frac{m - m_0}{p}\right] + \left[\frac{m_0 + n - m}{p}\right] + \left[\frac{m - m_0}{p^2}\right] + \left[\frac{m_0 + n - m}{p^2}\right] + \ldots < \frac{n}{p-1} \leqslant n.$$

Thus $|A'(m)|_p < p^{-n}$ and hence

$$|a_i|_p < p^n \max_{m_0 \leqslant m \leqslant m_0+n} |F(m)|_p \qquad (i = 0, 1, \ldots, n)$$

from which the assertion (63) follows.

§3. PRELIMINARY REMARKS

We now turn to our main objective, i.e. proving that, for any $w > n + 1$, the inequality

$$|F(\omega)|_p < h^{-w}, \quad h = h(F) \tag{64}$$

is, for almost all $\omega \in Q_p$, satisfied by only a finite number of polynomials F of degree n, with integer coefficients. First we show that the function $w_n(\omega)$, defined as the supremum of the set of those ω for which the inequality (64) has infinitely many solutions, is equal to a constant almost everywhere. Then we reduce the problem to the irreducible polynomials contained in the set

$$\mathbf{F}_n = \bigcup_{h=1}^{\infty} \mathbf{F}_n(h), \tag{65}$$

and from there on we shall follow the scheme of the proof as given in Part I.

If ω is a transcendental or algebraic p-adic number of a degree not exceeding n and if (64) has infinitely many polynomials $F(x)$, with integer coefficients, as solutions, then (64) is also satisfied by infinitely many primitive polynomials (with integer coefficients) provided that $w \geq 1$. This is evident since, if $F_1(x)$ is a primitive polynomial such that $F(x) = aF_1(x)$, then

$$h(F) = |a| h(F_1), \quad |a|_p \geq |a|^{-1}.$$

Therefore we shall henceforth consider primitive polynomials without needing any further elaboration.

Furthermore, let $\tilde{w}_n(\omega)$ denote the supremum of the set of those w for which the inequality (64) is solved by infinitely many irreducible polynomials with degrees not exceeding n.

By the same argument as in §5 of Chapter 1, Part I, the equation

$$\tilde{w}_n(\omega) = w_n(\omega) \qquad (n = 1, 2, \ldots) \tag{66}$$

may be derived easily.

These remarks will be useful later on, but for the moment we will continue to discuss the inequality (64) in relation to the set of all polynomials, with integer coefficients, of degree less than or equal to n.

Let $\mathrm{M}_n(w)$ be the set of those numbers $\omega \in Q_p$ for which there exists a constant $c = c(\omega, n, w)$ such that the inequality

$$|F(\omega)|_p < c h_F^{-w} \tag{67}$$

is satisfied by infinitely many polynomials F, with integer coefficients, with degrees not exceeding n. Then, if a is a rational integer, and if ω_1 is any one of the numbers $\omega + a$, $1/\omega$, or $a\omega$, then the inequality

$$|G(\omega_1)|_p < c_1 h_G^{-w} \tag{68}$$

formulated in analogy with (67), is satisfied by infinitely many polynomials G with degrees not exceeding n.

Indeed, the polynomials $G(x) = F(x - a)$, $x^n F(1/x)$, and $a^n F(x/a)$ are solutions of (68) for $\omega_1 = \omega + a$, $\omega_1 = 1/\omega$ and $\omega_1 = a\omega$, respectively. Thus either both or neither of the two numbers ω and ω_1 belong to the set $M_n(w)$. Hence it follows that, for any rational number r and any $\omega \in M_n(w)$, the number $\omega + r$ does also belong to this set.

In correspondence with §8 of the preceding chapter we now define the set $\sigma(F) = \sigma(F, w, c)$ of all numbers $\omega \in Q_p$ for which (67) holds with given F, w, and c. Furthermore, let

$$\Sigma_h(w, c) = \bigcup_{h(F) \leqslant h} \sigma(F, w, c), \tag{69}$$

where the union is taken with respect to all polynomials with integer coefficients whose degree and height do not exceed n and h, respectively. Finally, we define

$$\Sigma(w) = \bigcup_{c>0} \bigcap_{h=1}^{\infty} \Sigma_h(w, c), \tag{70}$$

where c runs through all positive real numbers. Clearly, $\Sigma(w)$ is the set of those $\omega \in Q_p$ for which there exists a constant c such that the inequality (67) has infinitely many polynomials F as solutions, i.e. $\Sigma(w) = M_n(w)$. Indeed, if $\omega \in \Sigma(w)$, then there exists a constant c such that

$$\omega \in \bigcap_{h=1}^{\infty} \Sigma_h(w, c).$$

Consequently, there exist infinitely many positive integers h for which $\omega \in \Sigma_h(w, c)$, and thus it follows from (69) that there exist infinitely many polynomials F with $\omega \in \sigma(F, w, c)$, i.e. polynomials which satisfy (67). Hence we do in fact have

$$\Sigma(w) = M_n(w). \tag{71}$$

By Lemma 12 of the preceding chapter the set $\sigma(F, w, c)$ is measurable. Therefore, (69), (70), and (71) imply the measurability of the set $M_n(w)$ for any w if we use the fact that the measurable sets form a σ-field.

If the set $M = M_n(w)$ has positive measure then its density $d(M)$ is equal to one by Lemma 6 of the preceding chapter. If M_0 is the set of rational numbers, Q an arbitrary circle with center 0, then it follows from Lemma 7 of that chapter that in this case

$$\frac{\mu\{(M + M_0) \cap \Omega\}}{\mu\Omega} = 1. \tag{72}$$

But since, for any $\omega \in M$, and any rational r, the number $\omega + r$ also belongs to M, we have $M + M_0 = M$. Consequently, (72) amounts to the assertion that, if $\mu M_n(w) > 0$, then

$$\frac{\mu\{M_n(w) \cap \Omega\}}{\mu\Omega} = 1$$

for any Ω. Thus we may conclude that, for any circle Ω and for any real number w,

$$\varphi(w) = \frac{\mu\{M_n(w) \cap \Omega\}}{\mu\Omega} = 1 \quad \text{or} \quad 0.$$

Clearly, if $w \leq w_1$, then $M_n(w) \supseteq M_n(w_1)$ and hence $\phi(w) \geq \phi(w_1)$. It follows easily from Lemma 3 that $\phi(n)$ is positive; hence $\phi(n) = 1$. But by the inequality (13′) of the Introduction there exists a w_0 with $\phi(w_0) = 0$. Therefore, there exists a constant w_n such that

$$\varphi(w) = \begin{cases} 1, & \text{if} \quad w < w_n, \\ 0, & \text{if} \quad w > w_n, \end{cases}$$

and thus

$$\frac{\mu\{M_n(w) \cap \Omega\}}{\mu\Omega} = \begin{cases} 1, & \text{if} \quad w < w_n, \\ 0, & \text{if} \quad w > w_n. \end{cases}$$

Hence it follows that, for any $\epsilon > 0$, the inequality (67) has for almost all $\omega \in Q_p$ infinitely many solutions (with a suitable constant $c = c(\omega)$) if we let $w = w_n - \epsilon$. On the other hand, this inequality has for almost all $\omega \in Q_p$ at most finitely many solutions (for any choice of c) if we let $w = w_n + \epsilon$. Thus it is evident that $w_n(\omega)$ is the supremum of the set of those $\omega \in Q_p$ for which (67) has infinitely many solutions (with a suitable c). Thus, $w_n(\omega) = w_n$ for almost all $\omega \in Q_p$.

Lemma 9. *There exist numbers* w_n *such that*

$$w_n(\omega) = w_n \quad (n = 1, 2, \ldots)$$

for almost all $\omega \in Q_p$.

§4. REDUCTION TO THE POLYNOMIALS FROM F_n

Let $w < w_n$, with w_n defined as in the preceding section. Allowing only primitive and irreducible polynomials of degree less than or equal to n, with integer coefficients, as solutions of the inequality (64), we can easily show that there exists a set Ω_1 of positive measure with the property that (64) has, for all $w < w_n$, infinite solutions $F(x)$ which satisfy the additional restriction

$$\max(|a_0|, |a_1|, \ldots, |a_{n-1}|) \ll |a_n|. \tag{73}$$

The proof of this assertion would amount to a repetition of Part I, Chapter 1, §6, without essential modifications. Furthermore, it was shown there that, under the condition (73), there exists a positive integer $m_0 \le c(n)$ such that

$$|F(m)| > \max_{(k)} |F^{(k)}(m)| \quad (k = 1, 2, \ldots, n-1)$$

for all the numbers $m = m_0, m_0 + 1, \cdots$. But since the polynomial $F(x)$ remains primitive and irreducible under the substitutions $x \to x + l$ and $x \to 1/x$, Lemma 9 guarantees the existence of a positive integer m_1 with $m_0 \le m_1 \le m_0 + n$ and $F(m_1) \not\equiv 0 \pmod{p^n}$.

Thus we may perform the transition to polynomials with the additional restrictions (47) by subjecting the set Ω_1 to the transformation $\omega \to 1/(\omega - m_0)$, thus mapping it onto a set Ω_0, and at the same time replacing F by the polynomial $\overline{F}_{(m_0)}$.

These considerations furnish a proof of the following lemma.

Lemma 10. *For any* $w < w_n$ *there exists a measurable set* Ω_0 *of positive measure such that the inequality* (64) *is, for all* $\omega \in \Omega_0$, *satisfied by infinitely many primitive and irreducible polynomials of degree* n, *with integer coefficients and subject to the conditions* (47), *i.e. polynomials* $F \in \mathrm{F}_n^*$, *where* F_n^* *denotes the set of polynomials with the specified properties.*

It follows from Lemmas 4 and 5 of the preceding chapter that if a set Ω_0 is obtained by selecting a measurable subset of positive measure from a circle and subjecting it to a transformation of the form $\omega \to \omega + l$ or $\omega \to 1/\omega$, then Ω_0 is itself a measurable set of positive measure.

§5. THE SIMPLEST SPECIAL CASES

The equation

$$w_1 = 2 \tag{74}$$

can be proved without difficulty. Indeed, we have $w_n \geq n+1$ $(n = 1, 2, \cdots)$ by Lemma 3; hence, in particular, $w_1 \geq 2$. Thus it suffices to show that $w_1 \leq 2$, i.e. that if a number $\epsilon > 0$ is given, the inequality

$$|a_0 + a_1\omega|_p < h^{-2-\varepsilon}, \quad h = \max(a_0, a_1) \tag{75}$$

has, for almost all $\omega \in Q_p$, at most finitely many solutions. We may assume that the numbers a_0 and a_1 are relatively prime since we always have either $(a_0, p) = 1$ or $(a_1, p) = 1$. If we suppose that $|\omega|_p \leq 1$ we necessarily obtain $(a_1, p) = 1$, since in the case $|a_0|_p = 1$ the inequality (75) implies $|a_0 + a_1\omega|_p < 1$; but if $|a_1|_p < 1$, one has $|a_0|_p < |a_1\omega|_p$. Thus, for $h > 1$, the inequality (75) implies

$$\left|\omega + \frac{a_0}{a_1}\right|_p < h^{-2-\varepsilon}, \quad h = \max(|a_0|, |a_1|). \tag{76}$$

For given h we denote by $\mathrm{M}(h)$ the set of all $\omega \in Q_p$ for which (76) is true. Thus we obtain

$$\sum_{h=1}^{\infty} \mu\, \mathbf{M}(h) \ll \sum_{h=1}^{\infty} h^{-1-\varepsilon} < \infty$$

for any $\epsilon > 0$. It follows from Part I, Chapter 1, Lemma 12, that the inequality (75) has, for almost all $\omega \in Q_p$ with $|\omega|_p \leq 1$, at most finitely many pairs a_0, a_1 of integers as solutions. If $|\omega|_p > 1$, we apply the transformation $\omega \to 1/\omega$ and turn to the case which we have already considered. So we obtain $w_1 \leq 2$, and, consequently, the assertion (74).

It is somewhat more difficult to establish the equation

$$w_2 = 3. \tag{77}$$

In order to prove it we apply Lemmas 6 and 10. It suffices to show that the inequality

$$|F(\omega)|_p < h_F^{-3-\varepsilon} \tag{78}$$

is satisfied by at most finitely many polynomials $F(x) = a_0 + a_1x + a_2x^2$ with integer coefficients; furthermore we may suppose that these polynomials are irreducible and that they have the properties stated in (47), i.e. that

$$\max(|a_0|, |a_1|) \leqslant a_2 = h, \quad |h|_p > p^{-2} \tag{79}$$

is true for them. If $|\omega - \kappa_1|_p \leq |\omega - \kappa_2|_p$, where κ_1 and κ_2 are the roots of $F(x)$, then it follows from Lemma 6 that

$$|\omega - \varkappa_1|_p \ll \frac{|F(\omega)|_p}{|D(F)|_p^{1/2}} < h^{-3-\varepsilon}\, |D(F)|_p^{-1/2}.$$

Consequently, if we denote again by $\mathrm{M}(h)$ the set of all $\omega \in Q_p$ for which (78) holds with a given h, we have

$$\sum_{h=1}^{\infty} \mu\, \mathrm{M}(h) \ll \sum_{h=1}^{\infty} h^{-3-\varepsilon} \sum_{h(F)=h}' |D(F)|_p^{-\frac{1}{2}}, \tag{80}$$

where in the summation

$$\sum_{h(F)=h}' |D(F)|_p^{-\frac{1}{2}} \tag{81}$$

only those polynomials are admitted for which $D(F) \neq 0$ in addition to the condition (79). In order to obtain an upper bound for this summation we first estimate the number $N(D)$ of pairs of integers a_0, a_1, subject to the condition (79), which satisfy the equation

$$D(F) = a_1^2 - 4a_0h = D \tag{82}$$

for a fixed integral value D.

Trivially, (82) implies

$$a_1^2 \equiv D \pmod{h}; \tag{83}$$

furthermore, a_0 is uniquely determined if a_1 is known. Hence $N(D)$ cannot be larger than the number of solutions of the congruence (83).

Lemma 11. *Let D and h be integers and let $N_0(D, h)$ denote the number of solutions of the congruence*

$$x^2 \equiv D \pmod{h}. \tag{84}$$

Then

$$N_0(D, h) \leqslant 2^{\nu(h)} \tau(h)(D, h)^{\frac{1}{2}}, \tag{85}$$

where $\nu(h)$ is the number of different prime divisors of h; furthermore, $\tau(h)$ is the number of divisors of h and (D, h) is the greatest common divisor of D and h.

Proof. First we show that

$$N_0(D, q^\delta) \leqslant 2\delta q^{\min\left(\left[\frac{\alpha}{2}\right],\, \left[\frac{\delta}{2}\right]\right)}, \tag{86}$$

for any prime number q with $q^\alpha \parallel D$.

Indeed, let $D = D_1 q^\alpha$ with $(D_1, q) = 1$.

a) If $\alpha \geq \delta$ then the congruence

$$x^2 \equiv D \pmod{q^\delta} \tag{87}$$

implies

$$x^2 \equiv 0 \pmod{q^\delta}. \tag{88}$$

The last congruence is certainly satisfied by $x \equiv 0 \pmod{q^\delta}$. If it has some solution x with $x \not\equiv 0 \pmod{q^\delta}$ then we write $x = x_1 q^u$ with $(x_1, q) = 1$, $0 \leq u < \delta$. In this case it follows from (88) that $2u \geq \delta$. For each u there are no more than $q^{\delta - u} \leq q^{[\delta/2]}$ solutions. Consequently we have

$$N_0(D, q^\delta) \leqslant 1 + \left[\frac{\delta}{2}\right] q^{\left[\frac{\delta}{2}\right]} \leqslant \delta q^{\left[\frac{\delta}{2}\right]}. \tag{89}$$

b) If $\delta > \alpha$ then, by (87),

$$x^2 \equiv q^\alpha D_1 \pmod{q^\delta}. \tag{90}$$

If we again write $x = x_1 q^u$, where $(x_1, q) = 1$, then we must have $2u = \alpha$, since it follows from (90) that

$$|x^2 - q^\alpha D_1|_q < q^{-\delta}, \quad |x|_q^2 = q^{-\alpha}.$$

Thus we obtain from (90) the congruence $x_1^2 \equiv D_1 (\mathrm{mod}\, q^{\delta-\alpha})$, and x_1 is expressible in the form

$$x_1 = x_2 + sq^{\delta-\alpha}, \quad 0 \le s < q^{\delta-u}/q^{\delta-\alpha} = q^{\alpha/2},$$

where s does not assume more than two values. Hence

$$N_0(D,\, q^\delta) \leqslant 2q^{\left[\frac{\alpha}{2}\right]}. \tag{91}$$

Together, (89) and (91) imply (86).

It follows from well-known facts in the theory of congruences that

$$N_0(D,\, h) = \prod_{q|h} N_0(D,\, q^\delta)$$

(see, for example, [70]. Chapter IV, §5). Hence, by (86),

$$N_0(D,\, h) \leqslant \prod_{q|h} 2\delta\, q^{\min\left(\left[\frac{\alpha}{2}\right],\, \left[\frac{\delta}{2}\right]\right)}$$

$$\leqslant 2^{\nu(h)}\, \tau(h) \prod_{q|(D,\, h)} q^{\min\left(\frac{\alpha}{2},\, \frac{\delta}{2}\right)} = 2^{\nu(h)}\, \tau(h)\, (D,\, h)^{\frac{1}{2}},$$

which proves the relation (85).

We now use the fact (see, for example, [70], Chapter II, Problem 11c) that, given an $\epsilon > 0$, the inequality $2^{\nu(h)} \le \tau(h) << h^\epsilon$ holds. Hence (86) implies that $N_0(D,\, h) << h^\epsilon (D,\, h)^{1/2}$.

Returning to the equation (82) and the congruence (83), we find

$$N(D) \ll h^\varepsilon (D,\, h)^{\frac{1}{2}}. \tag{92}$$

Now we can obtain an upper bound for the summation (81). By (92) one has

$$\sum_{h(F)=h}{}' |D(F)|_p^{-\frac{1}{2}} \leqslant \sum_{1 \leqslant |D| \leqslant h^2} N(D) |D|_p^{-\frac{1}{2}}$$

$$\ll \sum_{1 \leqslant D \leqslant h^2} h^\varepsilon (D,\, h)^{\frac{1}{2}} |D|_p^{-\frac{1}{2}}.$$

In order to estimate the summation

$$S = \sum_{1 \leqslant D \leqslant h^2} (D,\, h)^{\frac{1}{2}} |D|_p^{-\frac{1}{2}}. \tag{93}$$

We apply the Cauchy-Schwarz-Bunjakovskiĭ Inequality. If we let

$$S_1 = \sum_{1 \leqslant D \ll h^2} (D, h), \quad S_2 = \sum_{1 \leqslant D \ll h^2} |D|_p^{-1},$$

this inequality tells us that

$$S \leqslant S_1^{\frac{1}{2}} S_2^{\frac{1}{2}}. \tag{94}$$

The summation S_1 is estimated as follows:

$$S_1 \leqslant \sum_{D} \sum_{\substack{d|D \\ d|h}} d \ll \sum_{d|h} d\,\frac{h^2}{d} = h^2 \tau(h).$$

In order to find an upper bound for S_2 we denote by N_k the number of values of D, $1 \leq D \ll h^2$, for which $|D|_p = p^{-k}$. Since $N_k \ll h^2/k^2$ and since k runs through values satisfying $0 \leq k \ll \ln h/\ln p$, this leads to the inequality

$$S_2 = \sum_{k} N_k p^k \ll \sum_{k} \frac{h^2}{p^k}\, p^k \ll h^2 \ln h.$$

From (94) we have $S \ll h^2 (\tau(h) \ln h)^{1/2} \ll h^{2+\epsilon_1}$, and consequently the summation (81) does not exceed $c(\epsilon_2) h^{2+\epsilon_2}$. Going back to the inequality (80) we obtain, choosing $\epsilon_2 < \epsilon$, the relation

$$\sum_{h=1}^{\infty} h^{-3-\varepsilon} \sum_{h(F)=h}{}' |D(F)|_p^{-\frac{1}{2}} \ll \sum_{h=1}^{\infty} h^{-1-(\varepsilon-\varepsilon_2)} < \infty.$$

Hence

$$\sum_{h=1}^{\infty} \mu\, \mathbf{M}(h) < \infty,$$

and this implies that the equation (78) is satisfied by at most finitely many polynomials F, with integer coefficients, with the property stated in (79).

In order to free ourselves of the restriction (79) we apply Lemma 10. If $w_2 > 3$ then we choose a w with $3 < w < w_2$ and we consider the set Ω_0 mentioned in the lemma. Since $w < w_2$ we have, under the conditions of the lemma, $\mu\Omega_0 > 0$, and on the set Ω_0 the inequality (78) (with $w = 3 + \epsilon$) is satisfied by infinitely many polynomials F subject to the condition (79).

We have proved that, for almost all $\omega \in Q_p$, the inequality (78) has at most finitely many solutions which satisfy the condition (79). Thus we have reached a contradiction. Hence the assumption $w_2 > 3$ is false, i.e. we have $w_2 \leq 3$, from which the assertion (77) follows.

§6. DECOMPOSITION INTO ϵ-CLASSES

From now on we assume that $n \geq 3$. We choose an arbitrary $\epsilon > 0$ and subdivide the roots of each polynomial $F \in \mathrm{F}_n^*$ into classes by a procedure analogous to the method used in Part I.

Let $F \in \mathrm{F}_n^*$; furthermore, let $\kappa = \kappa_1$ be a root of F and let the remaining roots be numbered in such a manner that

$$|\varkappa_1 - \varkappa_2|_p \leqslant |\varkappa_1 - \varkappa_3|_p \leqslant \ldots \leqslant |\varkappa_1 - \varkappa_k|_p \leqslant 1 \leqslant \ldots \leqslant |\varkappa_1 - \varkappa_n|_p. \tag{95}$$

Here we make the assumption that there exists a root κ_k of the polynomials for which $|\kappa_1 - \kappa_k|_p \leq 1$. We introduce an integer m and real numbers ρ_i by the equations

$$m = \left[\frac{n}{\varepsilon}\right] + 1, \qquad |\varkappa_1 - \varkappa_i|_p = h^{-\rho_i} \quad (i = 2, 3, \ldots, k). \tag{96}$$

Let the integers $r_2, r_3, \cdots, r_k$ be defined by the inequalities

$$\frac{r_i}{m} \leqslant \rho_i < \frac{r_i + 1}{m} \quad (i = 2, 3, \ldots, k). \tag{97}$$

Then it is evident that

$$h^{-\frac{r_i+1}{m}} < |\varkappa_1 - \varkappa_2|_p \leqslant h^{-\frac{r_i}{m}} \quad (i = 2, 3, \ldots, k). \tag{98}$$

By (95) we have $\rho_2 \geq \rho_3 \geq \cdots \geq \rho_k \geq 0$, and it follows from (97) that

$$r_i = [m\rho_i], \quad r_2 \geqslant r_3 \geqslant \ldots \geqslant r_k \geqslant 0. \tag{99}$$

Now we may associate with each root $\kappa = \kappa_1$ of a given polynomial $F \in \mathrm{F}_n^*$ a vector $\mathfrak{r} = (r_2, r_3, \cdots, r_k)$ with nonnegative components. All κ's corresponding to the same vector $\mathfrak{r}$ are grouped together into a class $K_\epsilon(\mathfrak{r}) = K(\mathfrak{r})$.

In addition to the inequalities (99), the components of any such a vector $\mathfrak{r}$ satisfy the following relation:

Lemma 12. *If a class $K(\mathfrak{r})$ contains infinitely many elements, then*

$$\sum_{j=2}^{k}(j-1)\frac{r_j}{m}\leqslant n-1. \tag{100}$$

Proof. Using the product expansion (51) for the discriminant of a polynomial $F\in \mathbf{F}_n^*$ with a root $\kappa\in K(\mathfrak{r})$, we find

$$|D(F)|_p^{\frac{1}{2}}\ll\prod_{1\leqslant i<j\leqslant k}|\varkappa_i-\varkappa_j|_p\leqslant\prod_{1<j\leqslant k}|\varkappa_1-\varkappa_j|_p^{j-1}$$

$$=h^{-\sum_{j=2}^{k}(j-1)\rho_j}$$

Since, on the other hand, $|D(F)|_p>>h^{-2(n-1)}$, this implies

$$\sum_{j=2}^{k}(j-1)\rho_j\leqslant n-1+\frac{c\,(n,\,p)}{\ln h},$$

and hence, by (52),

$$\sum_{j=2}^{k}(j-1)\frac{r_j}{m}\leqslant n-1+\frac{c\,(n,\,p)}{\ln h}. \tag{101}$$

If the class $K(\mathfrak{r})$ has infinitely many elements we may let h tend to infinity in the last inequality, thus obtaining the assertion (100).

It should be noted that the inequality (101) together with (99) shows that the number of different classes $K(\mathfrak{r})$ is not greater than a constant $c(n,\,\epsilon,\,p)$.

§7. REDUCTION TO THE ROOTS OF A FIXED CLASS

For each root κ of a polynomial $F\in\mathbf{F}_n^*$ we introduce the set

$$S(\varkappa)=\{\omega:|\omega-\varkappa|_p=\min_{(\varkappa')}|\omega-\varkappa'|_p\},$$

where κ' runs through all roots of F.

If the inequality (19) is satisfied by infinitely many polynomials $F\in\mathbf{F}_n^*$ then the same polynomials will also satisfy the system of conditions

$$\left.\begin{aligned}&|F(\omega)|_p<h_F^{-w},\\&\omega\in S(\varkappa),\ \varkappa\in F\end{aligned}\right\}. \tag{102}$$

As in Part I we consider this system corresponding to each κ. (Since the polynomial F is irreducible, it is uniquely determined by any one of its roots κ.)

Let K_1^n be the set of those roots κ of the polynomials $F \in \mathrm{F}_n^*$ for which all conjugates κ' of κ, $\kappa' \neq \kappa$, satisfy the inequality $|\kappa - \kappa'|_p > 1$.

Lemma 13. *If* $w > n + 1$ *then the system* (102) *has, for almost all* ω, *at most finitely many solutions* $\kappa \in K_1^n$.

Proof. We obtain

$$|F'(\omega)|_p = |h|_p \prod_{\varkappa' \neq \varkappa} |\varkappa - \varkappa'|_p > |h|_p > p^{-n}.$$

By Lemma 9 of the preceding chapter,

$$|\omega - \varkappa|_p \leqslant \frac{|F(\omega)|_p}{|F'(\varkappa)|_p} \ll h^{-w}.$$

But since F_n^* does not contain more than $(2h+1)^n$ polynomials of height h and since, for any $w > n+1$, one has

$$\sum_{h=1}^{\infty} (2h+1)^n h^{-w} < \infty,$$

this proves the lemma.

Consequently, continuing with the discussion of the system (102), we may assume that $\kappa \notin K_1^n$, i.e. that there exists a conjugate root κ' of κ such that $|\kappa - \kappa'|_p \leq 1$. This means that κ may be associated with a class $K(\mathfrak{r})$ in the sense of the preceding section. So the system (102) is decomposed in a natural way into a finite number of systems of the form

$$\left.\begin{aligned} &|F(\omega)|_p < h_F^{-w}, \\ &\omega \in S(\varkappa),\ \varkappa \in F, \\ &\varkappa \in K(\mathfrak{r}), \end{aligned}\right\} \tag{103}$$

which correspond to the different classes $K(\mathfrak{r})$.

We shall distinguish between two categories among the classes $K(\mathfrak{r})$. Let

$$s_1 = \frac{1}{m}(r_3 + r_4 + \ldots + r_k), \tag{104}$$

if $k \geq 3$, and $s_1 = 0$ if $k = 2$. Then we shall say that a class $K(\mathfrak{r})$ is of the first or second kind according as to whether or not

$$\frac{r_2}{m} \leqslant \frac{n - s_1}{2}.$$

Our further discussion of the system (58) will depend on whether the class $K(\mathfrak{r})$ is of the first or of the second kind.

§8. INESSENTIAL DOMAINS

In this and the following section the system (103) will be studied under the assumption that $K(\mathfrak{r})$ is a class of the first kind.

In the sequel we will consider p-adic numbers ω which belong to a fixed circle Ω with its center at zero. By $\mathbf{F}_n(h)$ we denote the set of polynomials $F \in \mathbf{F}_n^*$ with $h(F) = h$. Letting ϵ and m be defined as in §6, we introduce the numbers $\delta = 2(\epsilon + 1/m)$; and

$$w_0 = w_{n-1} + \delta, \quad \delta > 0. \tag{105}$$

For each polynomial $F \in \mathbf{F}_n^*(h)$ we define the set $\sigma(F) = \sigma(F, w_0)$ of all $\omega \in \Omega$ for which

$$|F(\omega)|_p \leqslant h^{-w_0}; \tag{106}$$

furthermore, we define the sets $\sigma_i(F) = \sigma_i(F, w_0)$ by the equations

$$\sigma_i(F) = \sigma(F) \cap S(\varkappa_i) \quad (i = 1, 2, \ldots, n),$$

where $S(\kappa_i)$ is the set defined at the beginning of the preceding section. Without loss of generality we may assume that the inequalities (95) are satisfied.

Lemma 14. *If $\kappa = \kappa_1$ belongs to a class $K(\mathfrak{r})$ of the first kind then there exists a constant $c(n, p)$ such that for $h > c(n, p)$ the domain $\sigma_1(F)$ is a circle in Q_p. Precisely,*

$$\sigma_1(F) = \left\{\omega \in Q_p : |\omega - \varkappa_1|_p \leqslant \frac{h^{-w_0}}{|F'(\varkappa_1)|_p}\right\} \tag{107}$$

Proof. By Lemma 3 we have $w_{n-1} \geq n > 2$. Thus it follows from (106) by Lemma 8 that (61) is satisfied for all $\omega \in \sigma_1(F)$. Consequently, if we can show that

$$|\omega - \varkappa_1|_p < |\varkappa_1 - \varkappa_2|_p \tag{108}$$

for all $\omega \in \sigma_1(F)$ then we may conclude by means of Lemma 11 of the preceding chapter that $\sigma_1(F)$ is indeed the circle (107).

We observe that by (96) and (103) one has

$$h^{-s-\varepsilon} \ll |F'(\varkappa_1)|_p \ll h^{-s}, \tag{109}$$

where $s = r_2/m + s_1$. By Lemma 9 of the preceding chapter we have the inequality

$$|\omega - \varkappa_1|_p \leqslant \left(\frac{|F(\omega)|_p}{|F'(\varkappa_1)|_p} |\varkappa_1 - \varkappa_2|_p \right)^{\frac{1}{2}}$$

for all $\omega \in \sigma_1(F)$, and thus, taking (106) and (109) into consideration, we obtain

$$|\omega - \varkappa_1|_p \ll h^{-\frac{1}{2}(w_0 - s_1 - \varepsilon)} \tag{110}$$

Since, by assumption, κ_1 belongs to a class $K(\mathfrak{r})$ of the first kind as defined in §7, and since $w_{n-1} \geq n$, we have

$$\frac{r_2 + 1}{m} \leqslant \frac{n - s_1}{2} + \frac{1}{m} < \frac{w_0 - s_1 - \varepsilon}{2},$$

using the relation $\delta > \epsilon + 2/m$. Thus it follows from (110) for $h > c(n, p)$ that

$$|\omega - \varkappa_1|_p < h^{-\frac{r_2+1}{m}} \leqslant |\varkappa_1 - \varkappa_2|_p,$$

and this is the inequality we needed in order to apply Lemma 11 of the preceding chapter.

Now we may introduce the concepts of essential and inessential domains $\sigma_i(F)$ and establish a result which shall be important later on. These concepts are introduced and investigated in exactly the same way as in Part I.

The domain $\sigma_1(F)$ associated with a root $\kappa = \kappa_1$ of a polynomial $F \in \mathbf{F}_n^*(h)$ is called essential if the set of those points of $\sigma_1(F)$ which belong to any set $\sigma(G)$ with $G \in \mathbf{F}_n^*(h)$, $G \neq F$, has measure less than $2/\mu\sigma_1(F)$; otherwise the domain $\sigma_1(F)$ is called inessential. Clearly, a domain $\sigma_1(F)$ is essential if and only if

$$\mu \left\{ \sigma_1(F) \cap \left(\bigcup_{\substack{G \in \mathbf{F}_n^*(h) \\ G \neq F}} \sigma(G) \right) \right\} < \frac{1}{2} \mu\sigma_1(F), \tag{111}$$

and it is inessential exactly in the case where

$$\mu \left\{ \sigma_1(F) \cap \left(\bigcup_{\substack{G \in \mathbf{F}_n^*(h) \\ G \neq F}} \sigma(G) \right) \right\} \geqslant \frac{1}{2} \mu\sigma_1(F). \tag{112}$$

We denote by $\Delta_n(h)$ the set of those points $\omega \in \Omega$ for which there exists at least one polynomial F_1, with integer coefficients, not identically zero, whose degree and height do not exceed $n-1$ and $2h$, respectively, and which satisfies the condition

$$|F_1(\omega)|_p \leqslant h^{-w_0}. \tag{113}$$

It should be noted that if

$$\omega \in \sigma(F) \cap \sigma(G) \tag{114}$$

is true for some pair F, G of different polynomials from $\mathbf{F}_n^*(h)$, then $\omega \in \Delta_n(h)$. Indeed, (114) is equivalent with the truth of the two inequalities

$$|F(\omega)|_p \leqslant h^{-w_0}, \quad |G(\omega)|_p \leqslant h^{-w_0},$$

and hence the polynomial $F_1 = F - G$ satisfies (113) and all the other specified conditions, since $h(F) = h(G) = h$.

Hence we have, for a fixed polynomial $F \in \mathbf{F}_n^*(h)$,

$$\bigcup_{\substack{G \in \mathbf{E}_n^*(h) \\ G \neq F}} \sigma_1(F) \cap \sigma(G) \subseteq \Delta_n(h)$$

and consequently

$$\sigma_1(F) \cap \Big(\bigcup_{\substack{G \in \mathbf{E}_n^*(h) \\ G \neq F}} \sigma(G)\Big) = \bigcup_{\substack{G \in \mathbf{E}_n^*(h) \\ G \neq F}} \sigma_1(F) \cap \sigma(G)$$

$$\subseteq \sigma_1(F) \cap \Delta_n(h).$$

By (112) any inessential domain satisfies the inequality

$$\mu\{\sigma_1(F) \cap \Delta_n(h)\} \geqslant \mu\Big\{\sigma_1(F) \cap \Big(\bigcup_{\substack{G \in \mathbf{F}_n^*(h) \\ G \neq F}} \sigma(G)\Big)\Big\}$$

and thus

$$\mu\{\sigma_1(F) \cap \Delta_n(h)\} \geqslant \frac{1}{2}\,\mu\sigma_1(F). \tag{115}$$

Now we define

$$\Delta_n^{h_0} = \bigcup_{h \geq h_0} \Delta_n(h) \quad (h_0 = 1, 2, \ldots). \tag{116}$$

It is obvious that $\Delta_n^{h_0+1} \subseteq \Delta_n^{h_0}$ $(h_0 = 1, 2, \cdots)$. Furthermore, we introduce the set Δ_n by

$$\Delta_n = \bigcap_{h_0=1}^{\infty} \Delta_n^{h_0} = \lim_{h_0 \to \infty} \Delta_n^{h_0}. \tag{117}$$

By definition, Δ_n consists of those and only those points which belong to infinitely many sets $\Delta_n^{h_0}$, i.e. to infinitely many sets $\Delta_n^h(h)$. This means that for each $\omega \in \Delta_n$ there exist infinitely many polynomials F_1 of degrees less than n which satisfy inequalities of the form (113) (with variable h) in such a way that the height of F_1 does not exceed the corresponding constant $2h$. In other words, for each $\omega \in \Delta_n$ there exist infinitely many polynomials F_1 of degree less than n for which

$$|F_1(\omega)|_p \ll h_1^{-w_0}, \quad h_1 = h(F_1)$$

holds. From (105) and the definition of w_{n-1} it follows that the set Δ_n has measure zero. Thus we have by (72), using the continuity of the measure μ,

$$\lim_{h_0 \to \infty} \mu\Delta_n^{h_0} = \mu(\lim_{h_0 \to \infty} \Delta_n^{h_0}) = \mu\Delta_n = 0.$$

Therefore, given any $\epsilon_0 > 0$, we have

$$\mu\Delta_n^{h_0} < \varepsilon_0, \tag{118}$$

for all $h > h_0(\epsilon_0)$.

Continuing now with the investigation of inessential domains, we observe that (115) implies

$$\mu(\sigma_1(F) \cap \Delta_n^{h_0}) \geq \mu\{\sigma_1(F) \cap \Delta_n(h)\} \geq \frac{1}{2}\mu\sigma_1(F), \tag{119}$$

where $h = h(F)$, $h \geq h_0$, since it follows from (116) that $\Delta_n(h) \subseteq \Delta_n^{h_0}$.

We denote by $\Lambda(h_0)$ the union of all inessential domains $\sigma_1(F)$

corresponding to roots $\kappa = \kappa_1$ which belong to classes $K(\mathfrak{r})$ of the first kind and for which $h(\kappa_1) \geq h_0$. By Lemma 14 the domains $\sigma_1(F)$ may be considered as circles in Q_p. Hence by (118) and (119), Lemma 8 of the preceding chapter is applicable to the system $\Lambda = \Lambda(h_0)$ of domains and to the set $\Delta = \Delta_n^{h_0}$. Thus we obtain $\mu\Lambda(h_0) < 2\epsilon_0$ for all $h_0 > h_0(\epsilon_0)$. Consequently,

$$\lim \mu\Lambda(h_0) = 0. \tag{120}$$

Finally, we let Λ_0 denote the set of those points $\omega \in \Omega$ which are contained in in infinitely many inessential domains $\sigma_1(F)$ corresponding to roots κ_1 which belong to classes of the first kind, i.e.

$$\Lambda_0 = \bigcap_{h_0=1}^{\infty} \Lambda(h_0) = \lim_{h_0 \to \infty} \Lambda(h_0).$$

Then, by (120),

$$\mu\Lambda_0 = \mu(\lim_{h_0 \to \infty} \Lambda(h_0)) = \lim_{h_0 \to \infty} \mu\Lambda(h_0) = 0.$$

Thus we have proved the following assertion:

Proposition 1. *If Λ_0 denotes the set of those points $\omega \in \Omega$ which belong to infinitely many inessential domains $\sigma_1(F)$ corresponding to roots κ_1 contained in classes $K(\mathfrak{r})$ of the first kind, then Λ_0 has measure zero.*

§9. ESSENTIAL DOMAINS

The following lemma can be proved by exactly the same arguments as were used in Part I, Chapter 2, §2, Lemma 22.

Lemma 15. *Let $\lambda > 0$ be a real number and let $N(h, \lambda)$ denote the number of polynomials $F \in \mathbf{F}_n^*(h)$ to which there corresponds at least one essential domain $\sigma_i(F)$ with $\mu\sigma_i(F) \geq \lambda$. Then*

$$N(h, \lambda) \leqslant \frac{2}{\lambda} \mu\Omega. \tag{121}$$

If, for a fixed $\omega \in \Omega_p$, the system (103) has infinitely many solutions and if $w \geq w_0$, then ω belongs to infinitely many domains $\sigma_1(F)$ which correspond to roots $\kappa = \kappa_1$ satisfying (103). By Proposition 1 proved in the preceding section we only have to consider the case where $\sigma_1(F)$ is an essential domain. Thus we make the transition to the system

$$\left.\begin{aligned} &|F(\omega)|_p < h_F^{-w}, \\ &\omega \in S(\varkappa), \quad \varkappa \in F, \\ &\varkappa \in K(\mathfrak{r}), \\ &\sigma_1(F) \text{ essential.} \end{aligned}\right| \tag{122}$$

Here the domain $\sigma_1(F)$ corresponds to the root $\kappa = \kappa_1$ of the polynomial F, and we maintain the previous assumption that $K(\mathfrak{r})$ is a class of the first kind.

Let $\mathbf{M}_1(h, \mathfrak{r})$ be the set of those $\omega \in \Omega$ which satisfy (122) for some κ with

$$h(\varkappa) = h(F) = h. \tag{123}$$

By Lemma 9 of the preceding chapter we have

$$|\omega - \varkappa|_p \leqslant \frac{|F(\omega)|_p}{|F'(\varkappa)|_p},$$

and hence, applying the inequality (109),

$$|\omega - \varkappa|_p \ll h^{-w+s+\varepsilon}. \tag{124}$$

Consequently, $\mathbf{M}_1(h, \mathfrak{r})$ is covered by the system of circles in Q_p defined by the inequalities (124) for the different roots κ which satisfy the system (122) and the condition (123). In order to find an estimate for the outer measure of $\mathbf{M}_1(h, \mathfrak{r})$ it suffices to give an upper bound for the number $N(h, \mathfrak{r})$ of the different roots κ which satisfy the system (122) and the condition (123). We will obtain such a bound by applying Lemmas 14 and 15.

By Lemma 14 the domain $\sigma_1(F)$ is a circle $C(\kappa_1, \mathfrak{r})$, where $\mathfrak{r}$ is the smallest number of the form p^{-k} (with integral k) not exceeding $h^{-w_0}|F'(\kappa_1)|_p^{-1}$. By (109) one has $r \gg h^{-w_0+s}$ and thus

$$r \gg h^{-w_0+s},$$

$$\mu\sigma_1(F) = \mu C(\varkappa_1, r) \gg h^{-w_0+s}. \tag{125}$$

We now apply Lemma 15 with $\lambda \gg h^{-w_0+s}$. By the inequalities (125) and (121) it follows that

$$N(h, \mathfrak{r}) \ll h^{w_0-s}. \tag{126}$$

Finally, by (124) and (126) one has

$$\mu\, \mathbf{M}_1(h, \mathfrak{r}) \ll N(h, \mathfrak{r})\, h^{-w+s+\varepsilon} \ll h^{-w+w_0+\varepsilon}.$$

If we choose $w > w_0 + 1 + \epsilon$, this implies

$$\sum_{h=1}^{\infty} \mu\, \mathbf{M}_1(h, \mathfrak{r}) < \infty.$$

Consequently, by Lemma 12 of Part I, Chapter 1, §3, the system (122) has, for almost all $\omega \in \Omega$, at most finitely many solutions $\kappa \in K(\mathfrak{r})$, provided that $w > w_0 + 1 + \epsilon$. Returning now to the system (103) and applying Proposition 1 of the preceding section, we obtain the following result:

Proposition 2. *For almost all $\omega \in \Omega$ the system (103) is satisfied by at most finitely many algebraic numbers κ belonging to a given class $K(\mathfrak{r})$ of the first kind provided that $w > w_0 + 1 + \epsilon$.*

§10. CLASSES OF THE SECOND KIND

In this section we investigate the system (103) under the assumption that $K(\mathfrak{r})$ is a class of the second kind. Then we have, by the definition given in §7,

$$\frac{r_2}{m} > \frac{n - s_1}{2}. \tag{127}$$

It follows from Lemma 12 that

$$\frac{r_2}{m} + 2s_1 = \frac{r_2}{m} + 2\left(\frac{r_3}{m} + \ldots + \frac{r_k}{m}\right) \leqslant n - 1.$$

Consequently, by (127),

$$\frac{n - s_1}{2} + 2s_1 < \frac{r_2}{m} + 2s_1 \leqslant n - 1, \quad s_1 < \frac{n-2}{3}.$$

Furthermore, it is evident that

$$2\frac{r_3}{m} + s_1 \leqslant 3s_1 < n - 2, \quad \frac{r_3}{m} < \frac{n - s_1}{2} - 1.$$

Thus we have always

$$\frac{r_2}{m} > \frac{n - s_1}{2} > \frac{r_3}{m}. \tag{128}$$

Now let $\widetilde{\mathbf{F}}_n^*(2^t)$ be the subset of $\mathbf{F}_n^*$ consisting of all polynomials F which satisfy the condition

$$2^{t-1} \leqslant h(F) < 2^t \quad (t = 1, 2, \ldots). \tag{129}$$

Let $\mathbf{M}_2(2^t, \mathfrak{r})$ be the set of those $\omega \in \Omega$ for which (103) holds with $K(\mathfrak{r})$

being of the second kind and with $F \in \tilde{\mathrm{F}}_n^*(2^t)$. Then we have, by (103) and by Lemma 9 of the preceding chapter,

$$|\omega - \varkappa|_p \leqslant \left(\frac{|F(\omega)|_p}{|F'(\varkappa_1)|_p} |\varkappa_1 - \varkappa_2|_p \right)^{\frac{1}{2}}$$

and thus, by (109),

$$|\omega - \varkappa|_p \ll H^{-\frac{1}{2}(w - s_1 - \varepsilon)} \tag{130}$$

Here and below we let $H = 2^{t-1}$. Again, as in the last section, we have to find an estimate on the number $N(2^t, \mathfrak{r})$ of those $\kappa \in K(\mathfrak{r})$ (or of those polynomials $F \in \tilde{\mathrm{F}}_n^*(2^t)$) for which the conditions (103) can possibly be fulfilled.

First we show that for $t \gg 1$ the set $\tilde{\mathrm{F}}_n^*(2^t)$ cannot contain any pair F_1, F_2 of polynomials such that two of their respective roots $\kappa_1^{(1)}$ and $\kappa_1^{(2)}$ both belong to the class $K(\mathfrak{r})$ under consideration and both satisfy a condition of the form

$$|\varkappa_1^{(1)} - \varkappa_1^{(2)}|_p < cH^{-\frac{n-s_1}{2}}, \tag{131}$$

where c is a positive constant which is independent of t. Indeed, if such roots $\kappa_1^{(1)}, \kappa_1^{(2)}$ did exist, then we would have, by (95), (98) and (131),

$$\begin{aligned} |\varkappa_i^{(1)} - \varkappa_j^{(2)}|_p &\leqslant \max\left(|\varkappa_i^{(1)} - \varkappa_1^{(1)}|_p,\ |\varkappa_j^{(2)} - \varkappa_1^{(2)}|_p,\ |\varkappa_1^{(1)} - \varkappa_1^{(2)}|_p\right) \\ &\leqslant \max\left(h_1^{-\frac{1}{m}r_i},\ h_2^{-\frac{1}{m}r_j},\ cH^{-\frac{n-s_1}{2}}\right) \\ &\leqslant \max\left(H^{-\frac{1}{m}r_{\max(i,j)}},\ cH^{-\frac{n-s_1}{2}}\right), \end{aligned}$$

where $h_1 = h(F_1)$, $h_2 = h(F_2)$; $i, j = 1, 2, \cdots, k$, and where the relations $h_1 \geq H$, $h_2 \geq H$, $r_i \geq 0$ $(i = 1, 2, \cdots, k)$ have been used. Now we obtain from (128) the inequalities

$$|\varkappa_i^{(1)} - \varkappa_j^{(2)}|_p \leqslant \begin{cases} (c + \xi_H) H^{-\frac{n-s_1}{2}}, & \text{if } \max(i, j) \leqslant 2, \\ (1+c) H^{-\frac{1}{m} r_{\max(i,j)}}, & \text{if } \max(i, j) \geqslant 3, \end{cases}$$

where the sequence

$$\xi_H = H^{-\left(\frac{r_2}{m} - \frac{n-s_1}{2}\right)} \to 0 \quad (t \to \infty)$$

converges to zero as t tends to infinity. Consequently, considering the resultant $R(F_1, F_2)$ of the polynomials F_1, F_2, we obtain

$$|R(F_1, F_2)|_p < c(n,p) \prod_{1 \leqslant i,j \leqslant k} |\varkappa_i^{(1)} - \varkappa_j^{(2)}|_p$$

$$< c(n, p)(c+\xi_H)^4 (1+c)^{k^2} H^{-4\left(\frac{n-s_1}{2}\right)} \prod_{\max(i,j)\geqslant 3} H^{-\frac{1}{m} r_{\max(i,j)}}$$

Clearly

$$\frac{1}{m} \sum_{\max(i,j)\geqslant 3} r_{\max(i,j)} = s_1 + \frac{2}{m}(2r_3 + 3r_4 + \dots + (k-1)r_k) \geqslant 5s_1$$

and if we substitute this into the last inequality, it reduces to

$$|R(F_1, F_2)|_p < c(n, p)(c+\xi_H)^4 (1+c)^{k^2} H^{-2n-3s_1}. \tag{132}$$

Consequently, since $s_1 \geq 0$, it follows that $|R(F_1, F_2)|_p < c(n, p)H^{-2n}$, but as $R(F_1, F_2)$ is a rational integer $\ll H^{2n}$, this is obviously impossible for $H \gg 1$ and for a suitable value $c(n, p)$. Thus the inequality (131) leads to a contradiction if the constant $c > 0$ is chosen sufficiently small.

Therefore we may conclude that a circle $C(\kappa, \mathfrak{r})$ with its center at a point $\kappa \in K(\mathfrak{r})$, where $K(\mathfrak{r})$ is of the second kind, and with its radius r equal to the smallest number of the form p^{-k} (with integral k) not exceeding $cH(n - s_1)/2$, can never contain a root $\kappa' \in K(\mathfrak{r})$ of any polynomial $F \in \tilde{\mathrm{F}}_n^*(2^t)$ other than the minimal polynomial of κ. Hence

$$\mu C(\varkappa, r) \gg H^{\frac{n-s_1}{2}}.$$

We cover each of the numbers $\kappa \in K(\mathfrak{r})$ under consideration by the circle $C(\kappa, r)$, taking into account that these circles for different κ's are mutually disjoint, and, applying (132), we find that

$$N(2^t, \mathfrak{r}) \ll H^{\frac{n-s_1}{2}}. \tag{133}$$

By (130) and (133)

$$\mu \mathbf{M}_2(2^t, \mathfrak{r}) \ll N(2^t, \mathfrak{r}) H^{-\frac{1}{2}(w-s_1-\varepsilon)} \ll H^{-\frac{1}{2}(w-n-\varepsilon)}$$

Since $H = 2^{t-1}$ $(t = 1, 2, \cdots)$, this implies

$$\sum_{t=1}^{\infty} \mu \mathbf{M}_2(2^t, \mathfrak{r}) \ll \sum_{t=1}^{\infty} 2^{-\frac{t}{2}(w-n-\varepsilon)} < \infty$$

for all $w > n + \epsilon$.

Thus an application of Lemma 12 of Part I, Chapter 1, §3, leads to the following assertion:

Proposition 3. *Let* $w > n + \epsilon$. *Then the system* (103) *is, for almost all* $\omega \in \Omega$, *satisfied by at most finitely many numbers* κ *belonging to a given class* $K(\mathfrak{r})$ *of the second kind.*

§11. CONCLUSION OF THE PROOF

It is now easy to establish the equations

$$w_n = n + 1 \quad (n = 1, 2, \ldots). \tag{134}$$

If we choose

$$w > \max(w_0 + 1 + \varepsilon, \quad n + \varepsilon) \tag{135}$$

then it follows from Propositions 1 and 2 that the system (58), has, if a class $K(\mathfrak{r})$ is fixed, at most finitely many solutions for almost all $\omega \in \Omega$. Since $w_0 = w_{n-1} + \delta > n$ by Lemma 3 and (105), the inequality (135) is equivalent to the relation

$$w > w_{n-1} + 1 + \varepsilon + \delta. \tag{136}$$

Finally, we can show by means of Lemma 13 and the argument of §7 that under the condition (136) the system (102) has, for almost all ω, at most finitely many polynomials $F \in \mathrm{F}_n^*$ as solutions, and this statement is also true with regard to the inequality (64). But since in (136) ϵ and δ may be chosen arbitrarily small, this leads to the conclusion that (64) has, whenever $w > w_{n-1} + 1$, only finitely many polynomials $F \in \mathrm{F}_n^*$ as solutions. Thus it follows by Lemma 10 that the equation (134) holds.

Indeed, we have $w_n \geq n + 1$, by Lemmas 3 and 9. However, if $w_n > w_{n-1} + 1$, then we can choose a number w with $w_n > w > w_{n-1} + 1$, and because of the first part of this inequality it follows from Lemma 10 that there exists a set Ω_0 of positive measure such that the inequality (64) is satisfied, for almost all $\omega \in \Omega_0$, by infinitely many polynomials $F \in \mathrm{F}_n^*$. If Ω is chosen as a circle containing Ω_0 then, since $w > w_{n-1} + 1$, the remark made above means that Ω_0 has measure zero. The contradiction thus obtained shows that the inequality $w_n > w_{n-1} + 1$ is impossible. Therefore we have

$$w_n \geqslant n+1 \quad (n=1,\ 2,\ \ldots), \tag{137}$$

and also

$$w_n \leqslant w_{n-1}+1 \quad (n=3,\ 4,\ \ldots). \tag{138}$$

In §5 we have shown that $w_1 = 2$ and $w_2 = 3$. Consequently, it follows from (138) that $w_n \le n+1$ $(n = 1, 2, \cdots)$, and together with (137) this implies (134). Thus we have proved the following assertion.

Theorem. *For $\omega \in Q_p$ let $w_n(\omega)$ denote the supremum of the set of all $w > 0$ for which the inequality*

$$|a_0 + a_1\omega + \ldots + a_n\omega^n|_p < h^{-w}, \quad h = \max(|a_0|,\ \ldots,\ |a_n|),$$

has infinitely many integral solutions $a_0, a_1, \cdots, a_n$. Then

$$w_n(\omega) = n+1 \quad (n=1,\ 2,\ \ldots)$$

for almost all $\omega \in Q_p$.

CHAPTER 3

FIELDS OF FORMAL POWER SERIES

§1. NOTATION

In the sequel we shall use the following symbols:

Let K be a Galois field (finite field),

p the characteristic of K,

$q = p^f$ the number of elements in K,

x a transcendental element over K,

$R = K[x]$ the ring of polynomials in x with coefficients from K,

$K(x)$ the quotient field of $K[x]$, i.e. the field of rational functions in x over K,

$K\langle x\rangle$ the field of formal power series in x^{-1} over K, i.e. the set of elements $\omega = \omega(x)$ of the form

$$\omega(x) = \sum_{s=l}^{\infty} \alpha_s x^{-s}, \quad \alpha_s \in K \quad (s = l, l+1, \ldots), \tag{139}$$

with addition, subtraction, multiplication and division defined by the well-known rules,

$|\omega|$ the valuation on $K\langle x\rangle$ defined by the following conditions:

$$|\omega| = \begin{cases} 0 \text{ if in (139) } \alpha_s = 0 \text{ for all indices } s, \\ q^{-l} \text{ if } \alpha_l \neq 0, \end{cases}$$

$P, Q, \cdots$ polynomials from $R = K\langle x\rangle$ (or from other rings which will be specified),

$h(P)$ the height of the polynomial

$$P(z) = a_0(x) + a_1(x) z + \ldots + a_n(x) z^n, \quad a_i(x) \in R, \tag{140}$$

defined by the equation

$$h(P) = \max(|a_0(x)|, \ |a_1(x)|, \ \ldots, \ |a_n(x)|).$$

All further symbols will be defined in the text when they occur.

§2. BASIC FACTS FROM THE "GEOMETRY OF NUMBERS"

Let $\widetilde{K}$ be a field which is complete under a non-archimedean valuation, and let $\widetilde{K}_n$ be an n-dimensional vector space over $\widetilde{K}$. Mahler [46] introduced the concepts of convex sets and convex bodies in K_n, and he obtained analogs of Minkowski's well-known theorems on successive minima [47] and on linear forms ([6], Appendix B).

A real function $F(\bar{\omega})$ on $\widetilde{K}_n$ will be called metrical, if it satisfies the following conditions:

1) $F(0) = 0$; $F(\bar{\omega}) > 0$ if $\bar{\omega} \neq 0$,
2) $F(\alpha\bar{\omega}) = |\alpha| F(\bar{\omega})$ for all $\alpha \in \widetilde{K}$,
3) $F(\bar{\omega}_1 - \bar{\omega}_2) \leq \max(F(\bar{\omega}_1), F(\bar{\omega}_2))$.

For any real τ and any metric function F we define a convex body as the set $T(\tau)$ in K_n consisting of all points $\bar{\omega}$ which satisfy the inequality $F(\bar{\omega}) < \tau$. In particular, if $A = (\alpha_{ij})_{i,j=1,2,\cdots,n}$ is a nonsingular matrix over $\widetilde{K}$ and if

$$L_j(\bar{\omega}) = \sum_{i=1}^{n} \alpha_{ij}\omega_i \quad (j = 1, 2, \ldots, n), \tag{141}$$

then, for any n-tuple $c_1, \cdots, c_n$ of positive real numbers, the function

$$F(\bar{\omega}) = \max_{(j)} c_j^{-1} |L_j(\bar{\omega})| \tag{142}$$

is a metric function on $\widetilde{K}_n$. The convex bodies associated with such a function are called parallelepipeds. Mahler showed that, if the valuation in $\widetilde{K}$ is discrete, then a convex body in $\widetilde{K}_n$ may be approximated by parallelepipeds to any degree of accuracy. Furthermore, Mahler introduced the concept of the volume of a convex body (as some constant which is invariant under unimodular transformations of $\widetilde{K}_n$). In the case where $\widetilde{K} = K\langle x\rangle$ is a field of formal power series he proved an analog to Minkowski's theorem on convex bodies. Here the polynomials of the ring $R = K[x]$ play the role of the lattice points. In particular, the body $T(1)$ associated with the metric function (142) has the volume

$$V = V\{T(1)\} = \frac{c_1 c_2 \ldots c_n}{|\det(\alpha_{ij})|}. \tag{143}$$

It follows from Mahler's theorem on successive minima that the body $T(1)$ contains a nontrivial lattice point, i.e. a point $\mathbf{a} = (a_1, a_2, \cdots, a_n) \in T(1)$, $\mathbf{a} \neq (0)$, $a_i \in R$ $(i = 1, 2, \cdots, n)$, provided that $V > 1$.

In view of (142) and (143) this result may be reformulated as follows:

Lemma 1. *Let a linear form (141) over the field $K\langle x\rangle$ and n positive numbers $c_1, c_2, \cdots, c_n$ be given for which $c_1 c_2 \cdots c_n > |\det(\alpha_{ij})| \neq 0$. Then the system of inequalities*

$$\left|\sum_{i=1}^{n} \alpha_{ij} a_i\right| < c_j \quad (j = 1, 2, \ldots, n) \tag{144}$$

has a nontrivial integral solution, i.e. there exist polynomials a_i $(i = 1, 2, \cdots, n)$ in the ring $K[x]$, not all of them identically zero, which satisfy the inequalities (144).

The field K in this lemma is arbitrary. In the case where K is a finite field, the result of the lemma and other results of Mahler may be obtained by applying the theory of the measure introduced in Chapter 1. We are not going to elaborate on this point.

Later on we shall deal with solutions $(a_0, a_1, \cdots, a_n)$ of the inequality

$$|a_0 + a_1\omega + \ldots + a_n\omega^n| < h^{-w}, \tag{145}$$

where $a_i \in R = K[x]$ $(i = 0, 1, \cdots, n)$, $\omega \in K\langle x\rangle$, $w > 0$, $h = \max(|a_0|, |a_1|, \cdots, |a_n|)$. Turning now to the study of this inequality, we first establish the following results with the help of Lemma 1.

Lemma 2. *Let H be an arbitrary number greater than one. Then, for any $\omega \in K\langle x\rangle$, the inequality*

$$|a_0 + a_1\omega + \ldots + a_n\omega^n| < 2H^{-n}, \quad \max_{(i)} |a_i| < H(1 + |\omega|)^n$$

has a nontrivial integral solution.

Proof. The assertion follows immediately from Lemma 1, applied to the system

$$\begin{aligned} |a_0 + a_1\omega + \ldots + a_n\omega^n| &< 2H^{-n}, \\ |a_1| &< H, \\ \cdots\cdots\cdots&\cdots\cdots \\ |a_n| &< H. \end{aligned}$$

Lemma 3. *Let $w_n(\omega)$ denote the supremum of the set of those w for which the inequality* (145) *has infinitely many solutions. Then*

$$w_n(\omega) \geqslant n \quad (n = 1, 2, \ldots). \tag{146}$$

Proof. The assertion can be derived easily from Lemma 2.

§3. LEMMAS ON POLYNOMIALS

In this section we consider polynomials $P(z)$ over the ring $R^{1/d} = K[\sqrt[d]{x}]$, where d is a power of the prime number p, i.e. of the characteristic of K). The reduction of the problem to polynomials over this ring will be discussed in §4.

We assume that the polynomial

$$P(z) = a_0 + a_1 z + \ldots + a_n z^n, \quad a_i \in R^{1/d}, \tag{147}$$

satisfies the condition

$$\max(|a_0|, |a_1|, \ldots, |a_{n-1}|) \leqslant |a_n|. \tag{148}$$

By $\mathbf{P}_n^d$ we denote the set of polynomials of the form (147) which are irreducible over $R^{1/d}$, have no multiple roots and satisfy the condition (148). For any fixed element $a \in R^{1/d}$ let $\mathbf{P}_n^d(a)$ be the subset of $\mathbf{P}_n^d$ consisting of all polynomials for which

$$a_n = a, \quad a \in R^{1/d} \tag{149}$$

In connection with a polynomial $P \in \mathbf{P}_n^d(a)$ we sometimes consider the polynomials

$$P_i(z) = P_0(z)(z - \varkappa_i)^{-1} \quad (i = 1, 2, \ldots, n), \tag{150}$$

where κ_i runs through the roots of the polynomial $P_0(z) = P(z)$.

Let K^* be the smallest field containing $K\langle x\rangle$ and the roots of all polynomials over $R^{1/d}$. Since the set $R^{1/d}$, and hence the set of polynomials over $R^{1/d}$, is denumerable, we can introduce in K^* a valuation by means of the construction described in Chapter 1, §6. This valuation extends the given valuation in $K\langle x\rangle$; in particular, it is defined for the roots κ of the polynomials $P(z)$ over $R^{1/d}$. For each $\alpha \in K^*$ we denote the value thus defined by $|\alpha|$.

Lemma 4. *Let $P(z) \in \mathbf{P}_n^d(a)$ and let κ_i be the roots of $P(z)$. Then*

$$\max_{(i)} |\varkappa_i| \leqslant 1 \quad (i = 1, 2, \ldots, n). \tag{151}$$

Furthermore, if the polynomial $P(z)$ of the form (147) *satisfies the condition $|a_n| > ch(P)$, then*

$$\max_{(i)} |\varkappa_i| \leqslant c^{-1} \quad (i = 1, 2, \ldots, n).$$

Proof. Let κ be a root of $P(z)$. Then $a_0 + a_1\kappa + \cdots + a_n\kappa^n = 0$, and hence

$$a\varkappa = -a_{n-1} - a_{n-2}\varkappa^{-1} - \ldots - a_0\varkappa^{-n+1}.$$

If $|\kappa| > 1$ then (148) implies

$$|\varkappa| \leqslant |a|^{-1} \max(|a_{n-1}|, |a_{n-2}\varkappa^{-1}|, \ldots, |a_0\varkappa^{-n+1}|)$$
$$\leqslant |a|^{-1} \max(|a_{n-1}|, |a_{n-2}|, \ldots, |a_0|) \leqslant 1.$$

Thus we have obtained the relation $|\kappa| \leq 1$, which contradicts our assumption. Consequently this relation is always true, and this proves the assertion (151).

The second assertion of the lemma is proved analogously.

Lemma 5. *Let $P = P_0 \in \mathbf{P}_n^d(a)$ have κ among its roots and let the polynomials $P_j(z)$, $j = 1, 2, \cdots, n$, defined as in* (150), *have the degrees n_j, respectively. Then their discriminants $D(P_j)$ satisfy the inequalities*

$$|P_j'(\varkappa)| \geqslant |a|^{-n_j+2} |D(P_j)|^{\frac{1}{2}} \quad (j = 0, 1, \ldots, n). \tag{152}$$

Proof. We expand $D(P_j)$ in the form

$$D(P_j) = a^{2n-2} \prod_{1 \leqslant k < l \leqslant n} (\varkappa_k - \varkappa_l)^2. \tag{153}$$

By Lemma 4 we have $|\kappa_k - \kappa_l| \leq \max(|\kappa_k|, |\kappa_l|) \leq 1$ $(k, l = 1, 2, \cdots, n_j)$ and consequently

$$|D(P_j)| = |a|^2 \prod_{k \neq i} |\varkappa_i - \varkappa_k|^2 |a|^{2n_j-4} \prod_{\substack{1 \leqslant k < l \leqslant n_j \\ k,\, l \neq i}} |\varkappa_k - \varkappa_l|^2$$
$$\leqslant |P_j'(\varkappa_i)|^2 |a|^{2n_j-4} \tag{154}$$

from which the assertion (152) follows.

Lemma 6. *Let $\omega \in K\langle x\rangle$, $P \in \mathbf{P}_n^d(a)$, and let the roots be numbered in such a manner that*

$$|\omega - \varkappa_1| = \min_{(i)} |\omega - \varkappa_i| \quad (i = 1, 2, \ldots, n). \tag{155}$$

Then

$$|\omega - \varkappa_1| \leqslant \frac{|P(\omega)|}{|D(P)|^{1/2}} |a|^{n-2}. \tag{156}$$

Similarly, if

$$|\omega - \varkappa_2| = \min_{(i)} |\omega - \varkappa_i| \quad (i = 2, 3, \ldots, n),$$

then

$$|\omega - \varkappa_2| \leqslant \frac{|P_1(\omega)|}{|D(P_1)|^{1/2}} |a|^{n-3}. \tag{157}$$

Proof. We apply Lemma 9 of Chapter 1 and Lemma 5 to the polynomials $P(z)$ and $P_1(z)$.

Lemma 7. *If, under the conditions of Lemma 6,*

$$|P(\omega)| < |a|^{-w}, \quad w > 0 \tag{158}$$

and $n \geq 2$, *then*

$$|\omega - \varkappa_1| \leqslant |a|^{-1-w_1} |D(P)|^{-1/6}, \quad w_1 = \frac{2w - n}{3}. \tag{159}$$

Proof. The expansion (154) may (with $i = 1$, $j = 0$) be written in the form

$$|D(P)| = |P'(\varkappa_1)|^2 |D(P_1)|. \tag{160}$$

Let ρ be a real number to be determined later and suppose that $|D(P_1)| < |a|^{2n-4-2\rho}$. Then (160) implies

$$|P'(\varkappa_1)|^2 > |D(P)| \, |a|^{-2n+4+2\rho}$$

By Lemma 9 of Chapter 1 we have, using (158),

$$|\omega - \varkappa_1| \leqslant \frac{|P(\omega)|}{|P'(\varkappa_1)|} < |a|^{-w+n-2-\rho} |D(P)|^{-\frac{1}{2}}. \tag{161}$$

On the other hand, if $|D(P_1)| \geq |a|^{2n-4-2\rho}$, then we apply Lemma 6 to the polynomial $P_1(z)$, assuming again that

$$|\omega - \varkappa_2| = \min_{(i)} |\omega - \varkappa_i| \quad (i = 2, 3, \ldots, n).$$

Then by (157) we obtain the inequality

$$|\omega - \varkappa_2| \leqslant \frac{|P_1(\omega)|}{|D(P_1)|^{1/2}} |a|^{n-3} \leqslant \frac{|a|^{-w-1+\rho}}{|\omega - \varkappa_1|}.$$

Thus an application of (161) yields

$$|\omega - \varkappa_1| \leqslant \max\left(|a|^{-w+n-2-\rho}|D(P)|^{-\frac{1}{2}},\quad |a|^{-\frac{1}{2}(w+1-\rho)}\right).$$

Now we choose the number ρ so that the two expressions within the parentheses become equal to each other, and this leads to the assertion (159).

Lemma 8. *Let the roots of a polynomial $P(z) \in \mathbf{P}_n^d(a)$ be numbered in such a way that*

$$|\varkappa_1 - \varkappa_2| \leqslant |\varkappa_1 - \varkappa_3| \leqslant \ldots \leqslant |\varkappa_1 - \varkappa_n| \tag{162}$$

and let ω be an element of $K\langle x\rangle$ which satisfies the conditions (155) *and* (158) *with $w \geq n-1 \geq 2$. Then*

$$|\omega - \varkappa_1| \leqslant |\varkappa_1 - \varkappa_3| \tag{163}$$

Proof. Suppose that (163) is false, i.e. that $|\omega - \kappa_1| > |\kappa_1 - \kappa_3|$. Then (162) implies

$$\prod_{1 \leqslant i < j \leqslant 3} |\varkappa_i - \varkappa_j| < |\omega - \varkappa_1|^3. \tag{164}$$

By Lemma 7 we have, under the present conditions,

$$|\omega - \varkappa_1| \leqslant |a|^{-\frac{n+1}{3}} |D(P)|^{-\frac{1}{6}}.$$

But since the polynomial P has coefficients from the ring R and since $D(P) \neq 0$, it follows that $|D(P)| \geq 1$ and hence $|\omega - \kappa_1| \leq |a|^{-(n+1)/3}$. Together with (164) this yields

$$\prod_{1 \leqslant i < j \leqslant 3} |\varkappa_i - \varkappa_j| < |a|^{-n-1}. \tag{165}$$

Consequently, since (151) and (153) imply

$$1 \leqslant |D(P)| = |a|^{2n-2} \prod_{1 \leqslant i < j \leqslant n} |\varkappa_i - \varkappa_j|^2$$

$$\leqslant |a|^{2n-2} \prod_{1 \leqslant i < j \leqslant 3} |\varkappa_i - \varkappa_j|^2,$$

we obtain

$$\prod_{1\leq i<j\leq 3} |\varkappa_i - \varkappa_j| \geqslant |a|^{-n+1}.$$

This inequality is incompatible with (165), and thus the proof of the lemma is completed.

Lemma 9. *For any polynomial* $P(z)$,

$$\max_{(l)} |P(x^l)| \geqslant c(n) \max_{(i)} |a_i|,$$

where l *and* i *run through the values* $0, 1, \cdots, n$.

Proof. Let

$$Q(z) = (z-1)(z-x)\ldots(z-x^n),$$

and

$$Q_i(z) = \frac{Q(z)}{(z-x^i)Q'(x^i)} \quad (i = 0, 1, \ldots, n).$$

Then the polynomial $P(z)$ may be expressed in the form

$$P(z) = P(1)Q_0(z) + \ldots + P(x^n)Q_n(z). \tag{166}$$

If we use the notation

$$Q_i(z) = q_{i0}(x) + q_{i1}(x)z + \ldots + q_{in}(x)z^n \quad (i, j = 0, 1, \ldots, n)$$

we obtain the following identity by ordering the terms in (166) according to the powers of z:

$$a_j(x) = q_{0j}(x)P(1) + q_{1j}(x)P(x) + \ldots + q_{nj}(x)P(x^n). \tag{167}$$

Clearly $|q_{ij}(x)| \leq c(n)$, and hence, by (167),

$$\max_{(j)} |a_j(x)| \leqslant c(n) \max_{(l)} |P(x^l)| \quad (j, l = 0, 1 \ldots, n).$$

Lemma 10. *If* $P_1, P_2, \cdots, P_k$ *are polynomials from the ring* $R^{1/d}$ *then there exists a constant* $c > 0$, *depending only on their degrees, such that*

$$h(P_1P_2\ldots P_k) \geqslant ch(P_1)h(P_2)\ldots h(P_k).$$

Proof. It suffices to prove the lemma for pairs P_1, P_2 of polynomials, as the general case follows by induction on k.

We let $P(z) = P_1(z)P_2(z)$, $h(P) = h$. Denoting the degree of $P(z)$ with respect to z by m, we introduce the polynomials

$$P_{(j)}(z) = P(z + x^j), \tag{168}$$

and

$$\overline{P}(z) = z^m P\left(\frac{1}{z}\right). \tag{169}$$

Furthermore, the polynomial $\overline{P}_{(j)}(z)$ is defined by applying successively the transformations (168) and (169) to the polynomial P. Analogously, we also define the polynomials $P_{1(j)}$, $P_{2(j)}$, $\overline{P}_1$, $\overline{P}_2$, $\overline{P}_{1(j)}$ and $\overline{P}_{2(j)}$.

By Lemma 9, we have, for any exponent $l = 0, 1, \cdots, m$, the inequality

$$|P(x^l)| = \max_{(l')} |P(x^{l'})| > c(n)\,h, \tag{170}$$

where l' runs through the same range of values.

We consider the polynomial $Q = \overline{P}_{(l)} = \overline{P}_{1(l)} \overline{P}_{2(l)}$. From (170) we can conclude by Lemma 4 that under the valuation introduced above the roots of Q are bounded by $c(m)$. Consequently, the heights of the polynomials $\overline{P}_{(l)}(z)$, $\overline{P}_{1(l)}(z)$, $\overline{P}_{2(l)}(z)$ have the same order of magnitude as their leading coefficients, and thus the lemma is established in this case. The general case follows since the transformations (168) and (169) do not change the order of magnitude of the heights involved.

Actually, the lemma is true with the constant $c = 1$ and with the equality sign instead of $\geq$ (cf. [34], pp. 45–46).

§4. PRELIMINARY REMARKS

We want to show that, if $w > n$, the inequality

$$|P(\omega)| < h_P^{-w} \tag{171}$$

is, for almost all $\omega \in K\langle x\rangle$, satisfied by at most finitely many polynomials over $R = K[x]$ with degrees not exceeding n. It is evident that it suffices to allow only primitive polynomials as solutions of (171), i.e. polynomials for which the greatest common divisor of their coefficients is a constant from the field K. Let $w_n(\omega)$ be the supremum of the set of those $w > 0$ for which (171) has as solutions infinitely many polynomials $P(z)$ over R of degree less than or equal to n. Similarly, let $\widetilde{w}_n(\omega)$ denote the analogous supremum defined with the additional restriction that these polynomials be irreducible over R. Then we have

$$w_n(\omega) = \tilde{w}_n(\omega) \quad (n = 1, 2, \ldots). \tag{172}$$

A proof of this assertion may be given by means of Lemma 10 in analogy to the proof of the corresponding equations for real and complex numbers. Thus it suffices to investigate (171) for polynomials which are irreducible over R.

Let $P(z)$ be a polynomial of degree n, irreducible over R. Since the ring R has the finite characteristic $p \neq 0$, this does not exclude the possibility that $P(z)$ may have multiple roots. In that case the greatest common divisor $(P(z), P'(z))$ must equal $P(z)$, and hence $P'(z)$ must vanish identically. Consequently, there exists a polynomial $P_1(z)$ over R such that $P(z) = P_1(z^p)$. Then P_1 is irreducible over R, since otherwise $P(z)$ would also be reducible.

If $P_1(z)$ has multiple roots, then there exists a polynomial P_2 over R such that $P_1(z) = P_2(z^p)$, etc. It is easy to see that after a finite number of steps we obtain a polynomial $P_0(z)$, irreducible over R and without multiple roots, such that $P(z) = P_0(z^d)$ for some positive integer d. Then clearly d is of the form $d = p^l$ and $n = n_0 d$, where n_0 is the degree of the polynomial $P_0(z)$. Here n_0 and d are the reduced degree and the degree of inseparability of $P(z)$, respectively (see [81], Chapter II, §5).

Now let $P_0(z) = a_0 + a_1 z + \cdots + a_{n_0} z^{n_0}$, $a_i \in R$. Then we have

$$P(z) = P_0(z^d) = (\sqrt[d]{a_0} + \sqrt[a]{a_1}\, z + \ldots + \sqrt[d]{a_{n_0}}\, z^{n_0})^d = Q^d(z), \tag{173}$$

where $Q(z)$ is a uniquely determined polynomial since the root $\sqrt[d]{a}$ is determined for each $a \in R$ (see [78], Chapter V, §41). Furthermore it is evident that $\sqrt[d]{a} \in K[\sqrt[d]{x}]$, where $\sqrt[d]{x}$ is defined as a root of the polynomial $z^d - x$. Consequently, $Q(z)$ is a polynomial over $R^{1/d} = K[\sqrt[d]{x}]$.

Finally, let $\kappa_1, \kappa_2, \cdots, \kappa_{n_0}$ be the roots of the polynomial $P_0(z)$. Then

$$\begin{aligned} P_0(z^d) &= a_{n_0}(z^d - \kappa_1) \ldots (z^d - \kappa_{n_0}) \\ &= [\sqrt[d]{a_{n_0}}\,(z - \sqrt[d]{\kappa_1}) \ldots (z - \sqrt[d]{\kappa_{n_0}})]^d . \end{aligned}$$

Substituting this expression into (173) yields

$$Q(z) = \sqrt[d]{a_{n_0}}\,(z - \sqrt[d]{\kappa_1}) \ldots (z - \sqrt[d]{\kappa_{n_0}}),$$

which implies that $Q(z)$ does not have multiple roots.

With each polynomial $P(z)$ we associate the polynomial $Q_P(z) = Q(z)$ defined by the equation (173). It follows from this definition that

$$|P(\omega)| = |Q(\omega)|^d, \quad h(P) = h^d(Q),$$

and hence (171) is equivalent to the inequality

$$|Q(\omega)| < h_Q^{-w}, \tag{174}$$

where Q is a polynomial without multiple roots, irreducible over $R^{1/d}$ and of a degree not exceeding n/d. If (171) has infinitely many solutions for a given ω then there exists some d such that (174) also has infinitely many solutions. This means that, if $w^d_{n_1}(\omega)$ is the supremum of the set of those $w > 0$ for which (174) is satisfied by infinitely many polynomials $Q(z)$, without multiple roots, which are irreducible over $R^{1/d}$ and of degree not exceeding n, then

$$w_n(\omega) \leqslant \max w^d_{n_1}(\omega). \tag{175}$$

Here the maximum is taken with respect to all pairs n_1, d of positive integers, where d is of the form $d = p^l$, with $n_1 d \leq n$.

In connection with the inequality (175) our further efforts will be directed to proving the following result.

Proposition 1. *Let n be any positive integer and let $d = p^l$, $l = 0, 1, \cdots$. Then the inequality*

$$w^d_n(\omega) \leqslant nd \tag{176}$$

holds for almost all $\omega \in K\langle x\rangle$.

In turning to the proof of this proposition it is necessary to keep the following facts in mind.

Lemma 11. *There exist numbers w^d_n such that*

$$w^d_n(\omega) = w^d_n$$

for almost all $\omega \in K\langle x\rangle$ $(d = p^l,\ l = 0, 1, \cdots;\ n = 1, 2, \cdots)$.

A proof of this lemma would amount to a repetition of the arguments of §3 of the preceding chapter. We have to introduce sets $\mathbf{M}_n(w)$ and show that, if $\omega \in K\langle x\rangle$ belongs to the set $\mathbf{M}_n(\omega)$, it also contains the elements

$$\omega_1 = \omega + a, \quad a\omega, \quad \frac{1}{\omega} \quad (a \in R).$$

This implies that if $\omega \in \mathbf{M}_n(w)$ and $r \in K(x)$, then also $\omega + r \in \mathbf{M}_n(w)$. Since the elements of the field $K(x)$ are everywhere dense in $K\langle x\rangle$, the assertion just made follows by means of Lemma 6 of Chapter 1.

The measurability of the sets in question can be shown by a literal repetition of the set-theoretical consideration of §3 of the preceding chapter.

§5. REDUCTION TO THE POLYNOMIALS FROM $\mathbf{P}_n^d$

Lemma 12. *If $w < w_n^d$ then there exists a measurable set Ω_0 of positive measure such that, for almost all $\omega \in \Omega_0$, the inequality* (171) *is satisfied by infinitely many primitive polynomials $P(z)$ over $R^{1/d}$, without multiple roots, which are irreducible over $R^{1/d}$, have degrees not exceeding n and fulfill the condition* (148), *i.e. by polynomials $P(z) \in \mathbf{P}_n^d$.*

Proof. Let $\mathbf{P}_n(l)$ denote the set of all polynomials $P(z)$ of degree less than or equal to n which are irreducible over $R^{1/d}$ and do not have multiple roots, and for which, furthermore, l is the smallest nonnegative integer such that

$$|P(x^l)| = \max_{(j)} |P(x^j)| \quad (j = 0, 1, \ldots, n). \tag{177}$$

It is evident that each polynomial P with the indicated properties belongs to exactly one of the $n+1$ classes $\mathbf{P}_n(l)$ with $l = 0, 1, \cdots, n$.

Let Ω be an arbitrary circle in $K\langle x\rangle$ with center at zero. If $w < w_n^d$ then for almost all $\omega \in \Omega$ the inequality (171) is satisfied by infinitely many polynomials $P(z) \in R^{1/d}[z]$ with degrees not exceeding n. Let Ω_l be the set of those $\omega \in \Omega$ for which (171) has infinitely many solutions $P(z) \in \mathbf{P}_n(l)$. Obviously, Ω_l is measurable and

$$\bigcup_{l=0}^{n} \Omega_l \supseteq \Omega, \quad \sum_{l=0}^{n} \mu\Omega_l \geqslant \mu\Omega > 0.$$

Thus there exists an index l with $\mu\Omega_l \neq 0$. For this l, let Ω' be the set of points $\omega' \in K\langle x\rangle$ which are of the form $\omega' = (\omega - x^l)^{-1}$ with $\omega \in \Omega_l$. Since for each $\omega \in \Omega_l$ there exist infinitely many polynomials $P(z) \in \mathbf{P}_n(l)$ which satisfy (171), there are associated with these polynomials infinitely many polynomials $Q(z) = \overline{P_{(l)}}(z)$ (cf. (168) and (169)) which satisfy the inequality

$$|Q(\omega')| \ll h_Q^{-w} \tag{178}$$

Obviously, the leading coefficient of the polynomial $Q(z)$ is $P(x^l)$, and hence, by Lemma 9,

$$|P(x^l)| \gg h(P) \gg h(Q). \tag{179}$$

Finally, we map the set Ω' by the transformation

$$\omega' \to \omega'' = (\omega' - m_0)^{-1}; \quad \Omega' \to \Omega_0, \tag{180}$$

where m_0 is an integer to be defined below. At the same time we make the transition from the polynomial $Q(z)$ to $\overline{Q_{(m_0)}}(z)$.

By (179) and Lemma 4, for the roots $\kappa = \kappa^{(0)}$ of the polynomial $Q(z)$ and also for the roots $\kappa^{(k)}$ of its kth derivative $Q^{(k)}$ we have the relations

$$|\kappa^{(k)}| \leqslant c(n) = c \quad (k = 0, 1, \ldots, n-1). \tag{181}$$

Using again the notation $Q(z) = q_0 + q_1 z + \cdots + q_n z^n$, we let m be a positive integer satisfying $q^m > c$. Then $|x^m - \kappa| = q^m$ and thus

$$|Q(x^m)| = |q_n| |x^m - \kappa_1| \ldots |x^m - \kappa_n| \gg |q_n| q^{mn}. \tag{182}$$

It is evident from (181) that

$$|Q^{(k)}(x^m)| \leqslant h_Q q^{m(n-1)} \quad (k = 1, 2, \ldots, n-1). \tag{183}$$

We now define m_0 to be the smallest value of m for which $|q_n| q^{mn} \geq h_Q q^{m(n-1)}$. Since, by (179), $|q_n| > c_1 h(Q)$ with $c_1 = c(n)$, we obviously have $m_0 \leq c(n)$. Finally, (182) and (183) imply

$$|Q(x^{m_0})| > \max_{(k)} |Q^{(k)}(x^{m_0})| \quad (k = 1, 2, \ldots, n-1). \tag{184}$$

If we then consider the polynomial $\overline{Q_{(m_0)}}(z)$ instead of $Q(z)$, it follows from (184) that the new polynomial satisfies the condition (148). Thus the mapping (180) defines a set Ω_0 with the properties mentioned in Lemma 12.

§6. THE SIMPLEST SPECIAL CASES

Lemma 13. *For each integer $d = p^l$ with $l = 0, 1, \cdots$, the inequality*

$$w_1^d \leqslant d \quad (d = p^l, \quad l = 0, 1, \ldots) \tag{185}$$

holds.

Proof. It suffices to show that, given an $\epsilon > 0$, the inequality

$$|a_0 + a_1\omega| < H^{-(1+\varepsilon)d}, \quad H = \max(|a_0|, |a_1|), \tag{186}$$

has, for almost all $\omega \in K\langle x\rangle$, at most finitely many pairs a_0, a_1, of polynomials from $R^{1/d}$ as solutions. We may assume that $|\omega| \le 1$, since otherwise we only have to perform the transformation $\omega \longrightarrow 1/\omega$.

If $\omega \ne 0$ then $|a_1| = H$ by (186); thus

$$\left|\omega - \frac{a_0}{a_1}\right| < H^{-1-(1+\varepsilon)d}, \quad |a_0| \leqslant |a_1| = H. \tag{187}$$

Let $H = q^{h/d}$ and let $\mathbf{M}(H)$ be the set of those $\omega \in K\langle x\rangle$ for which (187) is true with some pair a_0, $a_1 \in R^{1/d}$, $a_1 \ne 0$. Obviously,

$$\mu\,\mathbf{M}(H) \ll H^{-1-(1+\varepsilon)d} M(H), \tag{188}$$

where $M(H)$ is the number of pairs of polynomials a_0, $a_1 \in R^{1/d}$ for which there exists an $\omega \in K\langle x\rangle$ such that (187) is true. It is clear that the condition $|a_1| = H$ amounts to the stipulation that a_1 be a polynomial in $\sqrt[d]{x}$ of degree h over K. The number of such polynomials is $q^h(q-1) < q^h = qH^d$. Concerning the polynomials a_0, their choice is subject not only to the condition $|a_n| \le H$ but also to the existence of an element $\omega \in K\langle x\rangle$ for which (187) holds. Indeed, letting $y = \sqrt[d]{x}$ and using the notation

$$\frac{1}{a_1} = \frac{1}{a_1(y)} = \alpha_h y^{-h} + \alpha_{h+1} y^{-h-1} + \cdots,$$

$$a_0 = a_0(y) = \beta_0 + \beta_1 y + \cdots + \beta_h y^h$$

we obtain from (187) the relation

$$\sum_{0 \leqslant j \leqslant h} \alpha_{s+j}\beta_j = 0 \quad (s = 0, 1, \ldots, h) \tag{189}$$

where s is not divisible by d since $\omega(x) = \omega(y^d)$. For a fixed a_1 the equation (189) imposes a linear condition on the coefficients β_j of the polynomial a_0. It should be noted that the determinant of the system (189) (with the β_j as unknowns, letting s run through the values $0, 1, \cdots, h$) does not vanish. This follows from the fact that the system (189) can have only the trivial solution because the conditions (189) are satisfied if and only if

$$\left\| \frac{a_0(y)}{a_1(y)} \right\| < q^{-h}, \quad \text{for} \quad \| a_0(y) \| < 1,$$

where the symbol $\| \ \|$ stands for the valuation relative to y. Thus, if the equation (189) is valid for some s which is not divisible by d, then there do not exist more than $(h+1)/d$ linearly independent coefficients of $a_0(y)$. Consequently, for each $a_1 \in R^{1/d}$ there do not exist more than $q^{(h+1)/d}$ polynomials $a_0 \in R^{1/d}$ for which (187) is possible with any $\omega \in K\langle x \rangle$.

Therefore we obtain the estimate $\mathbf{M}(H) \ll H^{d+1}$. Together with (188) this implies

$$\mu \mathbf{M}(H) \ll H^{-1-(1+\varepsilon)d+d+1} = H^{-\varepsilon d} = q^{-\varepsilon h}$$

and hence

$$\sum_H \mu \mathbf{M}(H) \ll \sum_{h=0}^{\infty} q^{-\varepsilon h} < \infty.$$

From this inequality the assertion (185) follows by Lemma 12 of Part I, Chapter 1.

By applying Lemmas 6 and 12 we can establish the inequalities

$$w_2^d \leqslant 2d \quad (d = p^l, \quad l = 0, 1, \ldots). \tag{190}$$

Let $\mathbf{P}_n^d(a)$ denote the set of those elements from $\mathbf{P}_n^d$ for which $a_n = a$, $a \in R^{1/d}$. Then, because of (156), the problem at hand reduces itself to finding an upper bound for the summation

$$\sum_{P \in \mathbf{P}_n^d(a)} |D(P)|^{-1/2}.$$

Now we proceed by the same method as in §5 of the preceding chapter. This requires that we obtain an estimate for the number of pairs of polynomials a_0, a_1 with $\max(|a_0|, |a_1|) \leq |a|$ satisfying the equation $a_1^2 - 4a_0 a = D$ for given polynomials a and D from the ring $R^{1/d}$. In order to obtain such an estimate it suffices to find a bound for the number of solutions of the congruence $a_1^2 \equiv D \pmod{a}$. This is accomplished by an argument which is entirely analogous to the approach used in §5 of the preceding chapter (cf. Chapter 2, Lemma 11). This type of analogy is due to the fact that the theory of divisibility in the ring $R^{1/d}$ is completely analogous to the theory of divisibility in the ring of rational numbers.

We shall not carry out in detail the indicated procedure, as the inequalities (190) are obtained by an argument which follows the general outline.

§7. DECOMPOSITION INTO ϵ-CLASSES

We choose an arbitrary $\epsilon > 0$ and consider it as fixed until the end of the proof.

Let $P \in \mathbf{P}_n^d$, let $\kappa = \kappa_1$ be a root of the polynomial P and let the remaining roots be numbered in such a way that

$$|\varkappa_1 - \varkappa_2| \leqslant |\varkappa_1 - \varkappa_3| \leqslant \ldots \leqslant |\varkappa_1 - \varkappa_n| \leqslant 1. \tag{191}$$

Then we have, by the assertion (151) of Lemma 4, the inequality $|\kappa_1 - \kappa_n| \leq 1$.

Again we introduce numbers m and ρ_i by letting

$$m = \left[\frac{n}{\varepsilon}\right] + 1, \quad |\varkappa_1 - \varkappa_i| = h^{-\rho_i} \quad (i = 2, 3, \ldots, n), \quad h = h(P) \tag{192}$$

and we define integers $r_2, r_3, \cdots, r_n$ by the inequalities

$$\frac{r_i}{m} \leqslant \rho_i < \frac{r_i + 1}{m} \quad (i = 2, 3, \ldots, n). \tag{193}$$

Then we have

$$h^{-\frac{r_i+1}{m}} < |\varkappa_1 - \varkappa_i| \leqslant h^{-\frac{r_i}{m}} \quad (i = 2, 3, \ldots, n) \tag{194}$$

and furthermore, by (191) and (192),

$$r_i = [m\rho_i], \quad r_2 \geqslant r_3 \geqslant \ldots \geqslant r_n \geqslant 0. \tag{195}$$

With each root $\kappa = \kappa_1$ of the polynomial $P \in \mathbf{P}_n$ we associate the vector $\mathbf{r} = (r_2, r_3, \cdots, r_n)$ with nonnegative integer components. All roots κ associated with a given vector $\mathbf{r}$ are grouped together into one class $K_\epsilon(\mathbf{r}) = K(\mathbf{r})$.

Lemma 14. *The integers r_j always satisfy the inequality*

$$\sum_{j=2}^{n} (j-1)\frac{r_j}{m} \leqslant n - 1. \tag{196}$$

Proof. We expand the discriminant of the polynomial P in the form (153), thus obtaining

$$|D(P)|^{\frac{1}{2}} \leqslant h^{n-1} \prod_{1<j\leqslant n} |\varkappa_1 - \varkappa_j|^{j-1} = h^{n-1-\sum_{j=2}^{n}(j-1)\rho_j}$$

Furthermore, since trivially $|D(P)| \geq 1$, it follows from (193) that

$$\sum_{j=2}^{n}(j-1)\frac{r_j}{m} \leqslant \sum_{j=2}^{n}(j-1)\frac{\rho_j}{m} \leqslant n-1$$

which proves the lemma.

For each root κ we define the set

$$S(\varkappa) = \{\omega\colon |\omega - \varkappa| = \min_{(\varkappa')}|\omega - \varkappa'|\},$$

where κ' runs through all roots of P.

Turning now to the study of the inequality (171) for polynomials $P \in \mathbf{P}_n^d$, we make the transition to the system

$$\left.\begin{aligned} &|P(\omega)| < h_P^{-w}, \\ &\omega \in S(\varkappa), \quad \varkappa \in P. \end{aligned}\right\} \tag{197}$$

We consider this as a system of conditions for κ, since, up to a factor from the field K, the polynomial P is uniquely determined by κ. Inasmuch as each κ belongs to some class $K(\mathbf{r})$, the system (197) can be decomposed into not more than $c(n, \epsilon)$ systems of the form

$$\left.\begin{aligned} &|P(\omega)| < h_P^{-w}, \\ &\omega \in S(\varkappa), \quad \varkappa \in P, \\ &\varkappa \in K(\mathbf{r}) \end{aligned}\right\} \tag{198}$$

corresponding to the different classes $K(\mathbf{r})$.

We subdivide the classes $K(\mathbf{r})$ into two categories. Let

$$s_1 = \frac{1}{m}(r_3 + r_4 + \ldots + r_n), \tag{199}$$

if $n \geq 3$, and $s_1 = 0$ if $n = 2$. Then we call $K(\mathbf{r})$ a class of the first or of the second kind according as to whether or not $r_2/m \leq (n - s_1)/2$.

§8. THE DOMAINS $\sigma_i(P)$

In the following three sections we assume the class $K(\mathfrak{r})$ in (198) to be of the first kind.

Let Ω be an arbitrary circle in $K\langle x\rangle$ with center at zero. Let $\epsilon > 0$ be the number defined in the preceding section; furthermore, let $\delta = 2(\epsilon + 1/m)$ and $w_0 = w_{n-1}^d + \delta$.

For each polynomial $P \in \mathbf{P}_n^d(a)$ we define the set $\sigma(P) = \sigma(P, w_0)$ of all $\omega \in \Omega$ for which

$$|P(\omega)| \leqslant h^{-w_0}, \quad h = |a| = h(P), \tag{200}$$

and furthermore we introduce sets $\sigma_i(P) = \sigma_i(P, w_0)$ by the equations

$$\sigma_i(P) = \sigma(P) \cap S(\varkappa_i) \quad (i = 1, 2, \ldots, n),$$

where $S(\kappa_i)$ is defined as in the preceding section. Without loss of generality we may assume in (198) that $\kappa = \kappa_1$ and that the convention (193) is observed.

Lemma 15. *If $\kappa = \kappa_1$ belongs to a class $K(\mathfrak{r})$ of the first kind then the domain $\sigma_1(P)$ is a circle in $K\langle x\rangle$; precisely,*

$$\sigma_1(P) = \left\{\omega \in K\langle x\rangle : |\omega - \varkappa_1| \leqslant \frac{h^{-w_0}}{|F'(\varkappa_1)|}\right\}. \tag{201}$$

Proof. First we observe that Lemma 3 implies the inequalities

$$w_n^d \geqslant n \cdot \quad (n = 1, 2, \ldots). \tag{202}$$

Indeed, let $y = \sqrt[d]{x}$ and consider the field $K\langle y\rangle$. In this field we define a valuation $\|\cdots\|$ by substituting y for x in the equation (139). Clearly, if $\omega \in K\langle x\rangle$, then $\omega = \omega(x) = \omega(y^d) \in K\langle y\rangle$. Furthermore,

$$\|\omega\| = |\omega|^d. \tag{203}$$

If, for some $w > 0$, the inequality

$$\|a_0(y) + a_1(y)\omega + \ldots + a_n(y)\omega^n\| < H^{-w},$$
$$H = \max(\|a_0(y)\|, \ldots, \|a_n(y)\|), \tag{204}$$

has a nontrivial solution, then it follows from (203) that this is also true for the inequality

$$|a_0(\sqrt[d]{x})+a_1(\sqrt[d]{x})\omega+\dots+a_n(\sqrt[d]{x})\omega^n|^d<H^{-w},$$

$$H=\max(|a_0|,\dots,|a_n|)^d,$$

and hence for the inequality

$$|a_0+a_1\omega+\dots+a_n\omega^n|<H_0^{-w},\quad a_i\in R^{1/d},$$

$$H_0=\max(|a_0|,\dots,|a_n|).$$

By Lemma 3 there exist infinitely many solutions of (204) with w arbitrarily close to n, and thus the relation (202) follows.

Now we turn to the assertion of Lemma 15. First let $n = 2$. Then we have, for any $\omega \in S(\kappa_1)$,

$$|\omega-\varkappa_1|=\begin{cases}|P(\omega)|:|P'(\varkappa_1)|, & |\omega-\varkappa_1|<|\varkappa_1-\varkappa_2|,\\ (|P(\omega)|:|a_2|)^{\frac{1}{2}}, & |\omega-\varkappa_1|\geqslant|\varkappa_1-\varkappa_2|.\end{cases}\tag{205}$$

In fact, if $|\omega-\kappa_1|<|\kappa_1-\kappa_2|$, then

$$|\omega-\varkappa_2|=|\omega-\varkappa_1+\varkappa_1-\varkappa_2|=|\varkappa_1-\varkappa_2|,\tag{206}$$

and hence

$$|\omega-\varkappa_1|=\frac{|\omega-\varkappa_1||\omega-\varkappa_2|}{|\varkappa_1-\varkappa_2|}=\frac{|P(\omega)|}{|P'(\varkappa_1)|}.$$

On the other hand, if $|\omega-\kappa_1|\geq|\kappa_1-\kappa_2|$, then

$$|\omega-\varkappa_2|=|\omega-\varkappa_1+\varkappa_1-\varkappa_2|$$

$$\leqslant\max(|\omega-\varkappa_1|,|\varkappa_1-\varkappa_2|)=|\omega-\varkappa_1|.$$

Since we are assuming that $\omega\in S(\kappa_1)$, we have $|\omega-\kappa_1|\leq|\omega-\kappa_2|$. Consequently, $|\omega-\kappa_1|=|\omega-\kappa_2|$ and thus

$$|\omega-\varkappa_1|=(|\omega-\varkappa_1||\omega-\varkappa_2|)^{\frac{1}{2}}=\left(\frac{|P(\omega)|}{|a_2|}\right)^{\frac{1}{2}}$$

which completes the proof of (205). We note that in the case $|\omega-\kappa_1|<|\kappa_1-\kappa_2|$ the condition $\omega\in S(\kappa_1)$ is automatically satisfied because of (206). This means that if we can establish the inequality

$$|\omega-\kappa_1|<|\kappa_1-\kappa_2|\tag{207}$$

for all $\omega\in\sigma_1(P)$, then the assertion of the lemma for $n = 2$ is proved

By (200) we have for all $\omega \in \sigma_1(P)$ the relation

$$|\omega - \varkappa_1| \leqslant h^{-\frac{1}{2}(w_0+1)} < |\varkappa_1 - \varkappa_2|, \quad h = h(P).$$

Here we have used the fact that (194) and the assumption about κ_1 belonging to a class of the first kind imply

$$|\varkappa_1 - \varkappa_2| > h^{-\frac{r_2+1}{m}} \geqslant h^{-1-\frac{1}{m}},$$

and that on the other hand, by (202),

$$\frac{1}{2}(w_0 + 1) \geqslant \frac{1}{2}(2 + \delta) = 1 + \varepsilon + \frac{1}{m} > 1 + \frac{1}{m}.$$

Now we come to the case $n \geq 3$. Since the condition (163) is satisfied by (202) and (200), Lemma 11 of Chapter 1 is applicable.

First we note that (194) implies the inequalities

$$h^{-s-\frac{n}{m}} < \prod_{i=2}^{n} |\varkappa_1 - \varkappa_i| \leqslant h^{-s},$$

$$h^{-1-s-\varepsilon} < |P'(\varkappa_1)| \leqslant h^{1-s}, \tag{208}$$

where $s = r_2/m + s_1$.

For any $\omega \in \sigma_1(P)$ we have, by Lemma 9 of Chapter 1,

$$|\omega - \varkappa_1| \leqslant \left(\frac{|P(\omega)|}{|P'(\varkappa_1)|}|\varkappa_1 - \varkappa_2|\right)^{\frac{1}{2}}.$$

Thus an application of (200) and (208) yields

$$|\omega - \varkappa_1| \leqslant h^{-\frac{1}{2}(w_0+1-s_1-\varepsilon)} \tag{209}$$

Furthermore, since κ_1 belongs to a class of the first kind, we obtain by means of (202) the relation

$$\frac{r_2+1}{m} \leqslant \frac{n-s_1}{2} + \frac{1}{m} \leqslant \frac{w_{n-1}^d + 1 - s_1}{2} + \frac{1}{m}$$

$$< \frac{w_0 + 1 - s_1 - \varepsilon}{2}.$$

Thus (207) follows from (209) and (194). This completes the proof of Lemma 15.

§9. INESSENTIAL DOMAINS

Let P be a polynomial from $\mathbf{P}_n^d(a)$ and κ one of its roots. The domain $\sigma_1(P)$ associated with the root $\kappa = \kappa_1$ will be called essential if the set of points in $\sigma_1(P)$ which belong to any system $\sigma(Q)$ with $Q \in \mathbf{P}_n^d(a)$, $Q \neq P$, has a measure less than $\frac{1}{2}\mu\sigma_1(P)$; otherwise we call the domain $\sigma_1(P)$ inessential. Obviously, the domain $\sigma_1(P)$ is essential if and only if

$$\mu\left\{\sigma_1(P) \cap \left(\bigcup_{\substack{Q \in \mathbf{P}_n^d(a) \\ Q \neq P}} \sigma(Q)\right)\right\} < \frac{1}{2}\mu\sigma_1(P),$$

and it is inessential if the converse inequality holds, i.e. if

$$\mu\left\{\sigma_1(P) \cap \left(\bigcup_{\substack{Q \in \mathbf{P}_n^d(a) \\ Q \neq P}} \sigma(Q)\right)\right\} \geqslant \frac{1}{2}\mu\sigma_1(P). \tag{210}$$

Let $\Delta_n(h)$ be the set of those points $\omega \in \Omega$ for which there exists in the ring $R^{1/d}[z]$ at least one polynomial P_1, not identically zero, whose degree and height do not exceed $n-1$ and h, respectively, and which satisfies the condition

$$|P_1(\omega)| \leqslant h^{-w}. \tag{211}$$

Obviously, if ω belongs to two overlapping sets $\sigma(P)$ and $\sigma(Q)$ associated with different polynomials P and Q, subject to the condition

$$P,\ Q \in \mathbf{P}_n^d(a), \quad |a| = h,$$

then $\omega \in \Delta_n(h)$, since in this case the polynomial $P_1 = P - Q$ satisfies the inequality (211).

By a repetition of the set-theoretical considerations of §8 of the preceding chapter it can be shown that, for any inessential domain, the inequality (210) implies

$$\mu\{\sigma_1(P) \cap \Delta_n(h)\} \geqslant \frac{1}{2}\mu\sigma_1(P).$$

From this inequality the following assertion may be derived by means of Lemma 8 of the preceding chapter together with Lemma 15:

Proposition 2. *Let Λ_0 denote the set of those points $\omega \in \Omega$ which belong* *to infinitely many inessential domains $\sigma_1(P)$ associated with roots κ_1 contained in classes $K(\mathfrak{r})$ of the first kind. Then Λ_0 has measure zero.*

§10. ESSENTIAL DOMAINS

We shall need the following analog of Part I, Chapter 2, §2, Lemma 22.

Lemma 16. *Let λ be an arbitrary positive number and $N(a, \lambda)$ the number of polynomials $P \in \mathbf{P}_n^d(a)$ which have at least one essential domain $\sigma_i(P)$ satisfying $\mu\sigma_i(P) \geq \lambda$. Then*

$$N(a, \lambda) \leqslant \frac{2}{\lambda}\mu\Omega. \tag{212}$$

Now we assume that in (198) the domain $\sigma_1(P)$ associated with $\kappa = \kappa_1$ is essential. This means that we are dealing with the following system of conditions:

$$\left.\begin{aligned} &|P(\omega)| < h_P^{-w}, \\ &\omega \in S(\varkappa), \quad \varkappa \in P, \\ &\varkappa \in K(\mathfrak{r}), \\ &\sigma_1(P) \text{ essential.} \end{aligned}\right\} \tag{213}$$

As before, we make the assumption that $K(\mathfrak{r})$ is a class of the first kind.

Let $\mathbf{M}_1(a, \mathfrak{r})$ be the set of those $\omega \in \Omega$ for which (213) holds with some κ whose minimal polynomial belongs to $\mathbf{P}_n^d(a)$. By Lemma 9 of Chapter 1 we have

$$|\omega - \varkappa_1| \leqslant \frac{|P(\omega)|}{|P'(\varkappa_1)|},$$

and hence, by (208),

$$|\omega - \varkappa_1| < h^{-w-1+s+\varepsilon} \tag{214}$$

Consequently, $\mathbf{M}_1(a, \mathfrak{r})$ can be covered by a system of circles in $K\langle x\rangle$ defined by the inequalities (214) for the different elements which satisfy the conditions of the system (213) and which have minimal polynomials $P \in \mathbf{P}_n^d(a)$. In order to obtain an upper estimate on the measure of $\mathbf{M}_1(a, \mathfrak{r})$ it suffices to find an upper bound for the number $N(a, \mathfrak{r})$ for the elements κ under consideration. Such a bound will be derived by an application of Lemma 16.

According to Lemma 15, the domain $\sigma_1(P)$ is a circle $C(\kappa_1, r)$ in $K\langle x\rangle$, where r is the smallest number for the form q^{-k} (with integral k) not exceeding $h^{-w_0}|P'(\kappa_1)|^{-1}$. By (208) we have $r \gg h^{-w_0+s-1}$ and thus

$$\mu\sigma_1(P) = \mu C(\varkappa_1, r) \gg h^{-w_0+s-1}$$

Now we apply Lemma 16 with $\lambda \gg h^{-w_0+s-1}$. Together with (212) this leads to the inequality $N(a, \mathfrak{r}) \ll h^{w_0-s+1}$.

Finally, by (214) and the estimate just obtained, we have

$$\mu M_1(a, \mathfrak{r}) \ll N(a, \mathfrak{r}) h^{-w-1+s+\varepsilon} \ll h^{-w+w_0+\varepsilon} \tag{215}$$

Let $M(h)$ be the number of polynomials $a \in R^{1/d}$ for which $|a| = h$. If $h = q^{k/d}$ then the condition $|a| = h$ means that $a = a(\sqrt[d]{x})$ is a polynomial of degree k in the variable $\sqrt[d]{x}$ over K. Hence

$$M(h) = (q-1)q^k. \tag{216}$$

Now by (215) and (216), we have the relation

$$\sum_a \mu M(a, \mathfrak{r}) \ll \sum_{k=0}^{\infty} q^k q^{-\frac{k}{d}(w-w_0-\varepsilon)} = \sum_{k=0}^{\infty} q^{-k\rho} < \infty,$$

provided that $\rho = (w - w_0 - \epsilon)/d - 1 > 0$, i.e. under the condition

$$w > w_0 + d + \varepsilon. \tag{217}$$

An application of Lemma 12 from Part I, Chapter 1, §3 furnishes the following result:

Proposition 3. *For almost all $\omega \in \Omega$ the system* (213) *under the condition* (217) *is satisfied by at most finitely many elements κ from a given class $K(\mathfrak{r})$ of the first kind.*

§11. CLASSES OF THE SECOND KIND

Now we assume that in the system (198) the class $K(\mathfrak{r})$ is of the second kind. Thus, by the definition given in §7, we have

$$\frac{r_2}{m} > \frac{n - s_1}{2}; \tag{218}$$

furthermore, by Lemma 14, $r_2/m + 2s_1 \leq n - 1$.

If $n \geq 3$ then (218) implies

$$\frac{r_2}{m} > \frac{n - s_1}{2} > \frac{r_3}{m}. \tag{219}$$

This inequality is also true for $n = 2$ since in that case we have $s_1 = r_3 = 0$.

Let $\mathbf{P}_n^d(h)$ be the set of those polynomials $P \in \mathbf{P}_n^d$ for which $|a_n| = h$. It follows from (198) by Lemma 9 of Chapter 1 that

$$|\omega - \varkappa_1| \leqslant \left(\frac{|P(\omega)|}{|P'(\varkappa_1)|} \, |\varkappa_1 - \varkappa_2| \right)^{\frac{1}{2}},$$

and thus, by (208),

$$|\omega - \varkappa_1| \leqslant h^{-\frac{1}{2}(w+1-s_1-\varepsilon)} \tag{220}$$

Now it remains to compute the number $N(h, \mathbf{r})$ of those polynomials $P \in \mathbf{P}_n^d(h)$ for which the condition (198) could possibly be satisfied.

We will show that in the set $\mathbf{P}_n^d(h)$ there is no pair P_1, P_2 of polynomials which are linearly independent over K and have roots $\kappa_1^{(1)}$ and $\kappa_1^{(2)}$, respectively, belonging to the class $K(\mathbf{r})$ under consideration and satisfying the inequality

$$|\varkappa_1^{(1)} - \varkappa_1^{(2)}| < h^{-\frac{n-s_1}{2}} \tag{221}$$

Indeed, otherwise it would follow from (191), (194) and (221) that

$$\begin{aligned} &|\varkappa_i^{(1)} - \varkappa_j^{(2)}| \\ &\leqslant \max(|\varkappa_i^{(1)} - \varkappa_1^{(1)}|, \, |\varkappa_j^{(2)} - \varkappa_1^{(2)}|, \, |\varkappa_1^{(1)} - \varkappa_1^{(2)}|) \\ &\leqslant \max\left(h^{-\frac{1}{m} r_i}, \, h^{-\frac{1}{m} r_j}, \, h^{-\frac{n-s_1}{2}}\right) \\ &= \max\left(h^{-\frac{1}{m} r_{\max(i,j)}}, \, h^{-\frac{n-s_1}{2}}\right). \end{aligned}$$

Thus, by (219), we would have

$$|\varkappa_i^{(1)} - \varkappa_j^{(2)}| \leqslant \begin{cases} h^{-\frac{n-s_1}{2}}, & \text{if} \quad \max(i, j) \leqslant 2, \\ h^{-\frac{1}{m} r_{\max(i,j)}}, & \text{if} \quad \max(i, j) \geqslant 3. \end{cases}$$

Therefore, again by (221), we would obtain the inequality

$$\begin{aligned} |R(P_1, P_2)| &< h^{2n} h^{-4\left(\frac{n-s_1}{2}\right)} \prod_{\max(i,j) \geqslant 3} h^{-\frac{1}{m} r_{\max(i,j)}} \\ &= h^{2s_1 - \frac{1}{m} \sum_{\max(i,j) \geqslant 3} r_{\max(i,j)}} \leqslant h^{-3s_1}. \end{aligned}$$

But since $s_1 \geq 0$, this would lead to the relation $|R(P_1, P_2)| < 1$. Consequently, $R(P_1, P_2) = 0$ and thus P_1, P_2 would differ only by a factor from K, contrary to the assumption made above.

Thus we have shown that for any $\kappa_1 \in K(\mathfrak{r})$, the circle $C(\kappa_1, r)$ (where r is the largest number not exceeding $h^{-(n-s_1)/2}$ of the form q^{-k}, with k integral) cannot contain a root $\kappa' \in K(\mathfrak{r})$ of any polynomial $P \in \mathbf{P}_n^d$ which is linearly independent of the polynomial defining κ_1.

Clearly, $\mu C(\kappa_1, r) \gg h^{-(n-s_1)/2}$, and hence we obtain the following estimate for the number $N(h, \mathfrak{r})$ of polynomials $P \in \mathbf{P}_n^d(h)$ for which (198) may have solutions:

$$N(h, \mathfrak{r}) \ll h^{\frac{1}{2}(n-s_1)} \tag{222}$$

Now an application of (220) and (222), letting $h = q^{k/d}$ $(k = 0, 1, \cdots)$, yields the following inequality for the measure $\mu \mathbf{M}_2(h, \mathfrak{r})$ of the set of those $\omega \in \Omega$ for which (198) is satisfied with $P \in \mathbf{P}_n^d(h)$ and with $K(\mathfrak{r})$ being of the second kind:

$$\mu \mathbf{M}_2(h, \mathfrak{r}) \ll h^{-\frac{1}{2}(w+1-s_1-\varepsilon)} h^{\frac{n-s_1}{2}} = h^{-\frac{1}{2}(w+1-n-\varepsilon)}$$

and consequently,

$$\sum_h \mu \mathbf{M}_2(h, \mathfrak{r}) \ll \sum_{k=0}^{\infty} q^{-\frac{k}{d}(w+1-n-\varepsilon)} < \infty,$$

provided that

$$w > n - 1 + \epsilon. \tag{223}$$

By making use of Part I, Chapter 12, §3, Lemma 12 we can complete the proof of the following assertion:

Proposition 4. *Under the condition* (223) *the system* (198) *has, for almost all* $\omega \in \Omega$, *at most finitely many solutions* κ *belonging to a given class* $K(\mathfrak{r})$ *of the second kind.*

§12. CONCLUSION OF THE PROOF

Evidently, it follows from Propositions 2–4 that for almost all $\omega \in \Omega$ the system (198) has at most finitely many solutions $\kappa \in K(\mathfrak{r})$, where $K(\mathfrak{r})$ is any fixed class, provided that $w > \max(w_0 + d + \epsilon, n - 1 + \epsilon) = w_0 + d + \epsilon$. (The last equation follows from the definition of w_0 and from (202).) Since the number of classes $K(\mathfrak{r})$ is bounded by some $c(n, \epsilon)$ and since the number ϵ can be chosen arbitrarily, it follows that the system (197), and hence also the inequality (171), have for almost all ω only finitely many polynomials $P \in \mathbf{P}_n^d$ as solutions, provided that

$$w > w_{n-1}^d + d \quad (n = 2, 3, \ldots). \tag{224}$$

Finally, (224) implies

$$w_n^d \leqslant w_{n-1}^d + d \quad (n = 2, 3, \ldots) \tag{225}$$

since otherwise we could choose a w with $w_n^d > w > w_{n-1}^d + d$, and then an application of Lemma 12 would lead to a contradiction to the result we have proved concerning the solvability of (171) by polynomials $P \in \mathbf{P}_n^d$.

From (225) we obtain by Lemma 13 the inequality $w_n^u \le nd$ $(n = 1, 2, \cdots; d = p^l, l = 0, 1, \cdots)$. Combined with Lemma 11 this completes the proof of Proposition 1 (inequality (176)). Now (175) leads to the conclusion that, for almost all $\omega \in K\langle x\rangle$, we have

$$w_n(\omega) \leqslant \max_{n_1 d \leqslant n} w_{n_1}^d \leqslant \max_{n_1 d \leqslant n} (n_1 d) = n,$$

where the maximum is taken with respect to all positive integers $n_1 d$ $(d = p^l, l = 0, 1, \cdots)$ satisfying $n_1 d \le n$. Since the opposite inequality was established by Lemma 3, we have thus proved the following result.

Theorem. *Let K be a finite field, $R = K[x]$ the ring of polynomials of x over K, $K\langle x\rangle$ the field of formal power series in x^{-1} over K, i.e. the field consisting of all series of the form*

$$\omega(x) = \sum_{s=l}^{\infty} \alpha_s x^{-s}, \quad \alpha_s \in K \quad (s = l, l+1, \ldots).$$

Let $|\omega|$ denote the valuation on $K\langle x\rangle$ defined by the equations

$$|\omega| = \begin{cases} 0, & \text{if } \alpha_s = 0 \quad (s = l, l+1, \ldots), \\ q^{-l}, & \text{if } \alpha_l \neq 0, \end{cases}$$

where q is the number of elements of the field K. Finally, let $w_n(\omega)$ denote the supremum of the set of those numbers w for which the inequality

$$|a_0 + a_1\omega + \ldots + a_n\omega^n| < h^{-w}, \quad a_i \in R,$$
$$h = \max(|a_0|, \ldots, |a_n|),$$

is satisfied by infinitely many systems $a_0, a_1, \cdots, a_n$ of polynomials from the ring R. Then the equations $w_n(\omega) = n$ $(n = 1, 2, \cdots)$ hold for almost all $\omega \in K\langle x\rangle$ (in the sense of the measure defined over that field).

SUPPLEMENTARY RESULTS AND REMARKS

A. REAL AND COMPLEX NUMBERS

Dirichlet series with "small denominators". Let $\omega_1, \omega_2, \cdots, \omega_n$ be real numbers, and let $w_n(\omega_1, \cdots, \omega_n)$ be the supremum of the set of those $w > 0$ for which there exist infinitely many solutions of the inequality

$$|a_0 + a_1\omega_1 + a_2\omega_2 + \ldots + a_n\omega_n| < h^{-w},$$

$$h = \max(|a_0|, |a_1|, \ldots, |a_n|).$$

We consider the series

$$S = \sum_{a_0, a_1, \ldots, a_n} h^{-s}|a_0 + a_1\omega_1 + \ldots + a_n\omega_n|^{-1},$$

$$h = \max(|a_0|, |a_1|, \ldots, |a_n|) \neq 0, \tag{1}$$

where the exponent s is a complex variable. We will show that the abscissa of absolute convergence (see, for example, [7], p. 62) of this series coincides with $w_n(\omega_1, \omega_2, \cdots \omega_n)$.

Indeed, assuming $w_n(\omega_1, \omega_2, \cdots, \omega_n) < \infty$, let

$$S_H = \sum_{a_0, \ldots, a_n} |a_0 + a_1\omega_1 + \ldots + a_n\omega_n|^{-1},$$

$$0 \neq \max(|a_0|, |a_1|, \ldots, |a_n|) \leqslant H. \tag{2}$$

The number N of terms in the summation (2) is $\ll H^{n+1}$. Let $\rho_1, \cdots, \rho_N$ with $0 < \rho_1 \leq \rho_2 \leq \cdots \leq \rho_N$ be the values of the expression $|a_0 + a_1\omega_1 + \cdots + a_n\omega_n|$, numbered in the order of increasing magnitude.

For a fixed $w > w_n(\omega_1, \cdots, \omega_n)$ it is clear from the definition of $w_n(\omega_1, \cdots, \omega_n)$ that $\rho_{k+1} - \rho_k \gg H^{-w}$ $(k = 1, \cdots, N-1)$. Indeed, if we use the notation

$$\rho_{k+1} = a_0^{(1)} + a_1^{(1)}\omega_1 + \ldots + a_n^{(1)}\omega_n,$$

$$\rho_k = a_0^{(0)} + a_1^{(0)}\omega_1 + \ldots + a_n^{(0)}\omega_n,$$

then

$$\rho_{k+1}-\rho_k = (a_0^{(1)} - a_0^{(0)}) + (a_1^{(1)} - a_1^{(0)})\,\omega_1 + \ldots + (a_n^{(1)} - a_n^{(0)})\,\omega_n,$$

$$\max\,(|a_0^{(1)} - a_0^{(0)}|, \ldots, |a_n^{(1)} - a_n^{(0)}|) \leqslant 2H$$

and thus the asserted inequalities follow. Consequently we have $\rho_k >> kH^{-w}$ $(k = 1, \cdots, N)$, and hence the summation (2) satisfies

$$S_H \ll \sum_{k\leqslant N} H^w \frac{1}{k} \ll H^w \sum_{k\leqslant N} \frac{1}{k} \ll H^w \ln H.$$

Finally, applying summation by parts to the initial series (1), we obtain

$$S \ll \sum_{H=1}^{\infty} S_H H^{-s-1} \ll \sum_{H=1}^{\infty} H^{w-s-1} \ln H < \infty,$$

provided that $s > w$. Consequently, the series (1) converges for any $s > w_n(\omega_1, \omega_2, \cdots, \omega_n)$.

Conversely, it follows from the definition of the quantity $w_n(\omega_1, \cdots, \omega_n)$ that in the case where $s \leq w < w_n(\omega_1, \cdots, \omega_n)$, there are infinitely many terms in the summation for which

$$h^{-s}\,|a_0 + a_1\omega_1 + \ldots + a_n\omega_n|^{-1} > h^{-s+w} \gg 1.$$

Hence the series (1) is divergent for any $s < w_n(\omega_1, \cdots, \omega_n)$. We wish to remark that series of the form (1) occur in connection with certain problems of mathematical physics (see, for example, V. I. Arnol′d [2]).

On another classification of the transcendental numbers.[1] Starting from the measure $w_n(\omega, H)$ of transcendency, it is possible to introduce a classification of the (real or complex) numbers which formally resembles Mahler's classification but possesses a number of features which do not prevail in the classification of Mahler.

We define

$$v(\omega, H) = \overline{\lim_{n\to\infty}} \frac{\ln\ln \dfrac{1}{w_n(\omega, H)}}{\ln n},$$

$$v(\omega) = \sup_{(H)} v(\omega, H) \quad (H = 1, 2, \ldots).$$

The quantity $v(w)$ will be called the order of the number ω. If this order is a finite number $v(\omega) = v$ then we let

---

1) Cf. [61].

$$t(\omega, H) = \overline{\lim} \frac{\ln \dfrac{1}{w_n(\omega, H)}}{n^v}$$

$$t(\omega) = \overline{\lim}_{H \to \infty} \frac{t(\omega, H)}{\ln H}.$$

We will call $t(\omega)$ the type of the number ω.

If $v(\omega) = \infty$ and if there exists an H such that $v(\omega, H) = \infty$, then let $H_0 = H_0(\omega)$ denote the smallest value of H for which this is true; if no such H exists, let $H_0 = \infty$.

Now we introduce classes of $\tilde{A}$-, $\tilde{S}$-, $\tilde{T}$- and $\tilde{U}$-numbers by the following scheme: We call ω

an $\tilde{A}$-number if $0 \leq v(\omega) < 1$ or if $v(\omega) = 1$ and $t(\omega) = 0$,
an $\tilde{S}$-number if $1 < v(\omega) < \infty$ or if $v(\omega) = 1$ and $t(\omega) > 0$,
a $\tilde{T}$-number if $v(\omega) = \infty$ and $H_0(\omega) = \infty$,
a $\tilde{U}$-number if $v(\omega) = \infty$ and $H_0(\omega) < \infty$.

It is easy to prove the following assertions:

(1) All algebraic numbers have orders not exceeding 1. If an algebraic number has order 1 then it is of type 0.

(2) All transcendental numbers have orders not less than 1. If a real transcendental number has order 1 then it is of a type not less than 1. If a complex transcendental number has order 1 then it is of a type not less than ½.

Indeed, let $\mathbf{P}_n(H)$ be the set of all polynomials with integer coefficients whose degree and height do not exceed n and H, respectively, and let $\omega_0 = \max(1, |\omega|)$.

Let ω be an algebraic number of degree m, denote the minimal polynomial of ω by $f(x)$, and its leading coefficient by a. Furthermore, let P be an element of $\mathbf{P}_n(H)$ with $P(\omega) \neq 0$. Then

$$1 \leqslant |R(P, f)| \leqslant |a|^n |P(\omega)| (\omega_1^n (n+1) H)^{m-1},$$

where $\omega_1 = \max |\omega'|$ as ω' runs through all roots of the polynomial $f(x)$. Thus the first assertion follows.

The second assertion follows from the fact that there exists, by Dirichlet's pigeonhole principle, a polynomial $P \in \mathbf{P}_n(H)$ which satisfies the condition

$$0 \neq |P(\omega)| < (n+1)\,\omega_0^n H^{-n},$$

or the condition

$$0 \neq |P(\omega)| < 2(n+1)\,\omega_0^n H^{-\frac{n+1}{2}}$$

for real and for complex values, respectively, the transcendental number ω.

Thus it follows from the first assertion that the algebraic numbers form the class of $\tilde{A}$-numbers.

According to results of N. I. Fel'dman [12] the numbers π and $\ln \alpha$ for algebraic $\alpha \neq 0, 1$ have orders not greater than 2 and 3, respectively;[1)] hence they are $\tilde{S}$-numbers.

The existence of $\tilde{U}$-numbers can be shown constructively in the manner of LeVeque [37].

Concerning $\tilde{T}$-numbers, their existence is as much a mystery as the existence of T-numbers in the sense of Mahler.

It can be shown without difficulty that almost all real or complex numbers belong to the class of $\tilde{S}$-numbers and have orders not exceeding 2. It seems likely to us that almost all of these numbers have order 1.

Finally, one may introduce still other classifications of the real or complex numbers (for example, by allowing the parameters n and H to change simultaneously) but the crux of the matter is that it is difficult to devise a classification which really classifies almost all numbers (in the sense that there is no set of positive measure whose elements are indistinguishable under the classification) and whose classes have certain invariance properties. Indeed, if K is any set of real numbers such that $\omega \in \mathrm{K}$ always implies $\omega + r \in \mathrm{K}$ for any rational r then either K or its complementary set is of measure zero (cf. Part I, Lemma 11). This feature leads to the fact that in Mahler's classification, where the classes are algebraically closed (cf. Schneider [54]) and thus possess a strong invariance property, the basic metric function $w_n(\omega)$ is almost everywhere constant. Hence there exists a set containing almost all numbers such that the classification does not distinguish between them.

---

1) In recent years Fel'dman sharpened these bounds to 1 and 2, respectively.

The classification we have defined above probably does not have trivial invariance properties, and even under the assumption that the order $v(\omega)$ is almost everywhere constant it would still be difficult to show this to be true also for the type $t(\omega)$.

Successive minima of polynomials.[1] Let the real number ω be transcendental or algebraic of degree greater than n and let $\mathbf{P}_n(H)$ be the set consisting of all polynomials with integer coefficients whose degree and height do not exceed n and H, respectively, and of all derivatives of such polynomials. We select in the set $\mathbf{P}_n(H)$ a polynomial P whose absolute value at the point ω is as small as possible. Then the polynomial P is uniquely determined up to the sign. Indeed, if two different polynomials P and P_1 both have the indicated minimal property then obviously $|P(\omega)| = |P_1(\omega)|$. Consequently, ω is a root of either the polynomial $P + P_1$ or of the polynomial $P - P_1$. Hence ω is an algebraic number of a degree not exceeding n, contrary to our assumption.

Clearly, the polynomial P which is minimal in the sense described at a point ω maintains this minimal property throughout some neighborhood (α, β) of the point ω. Then we call the polynomial P the minimal polynomial of the set $\mathbf{P}_n(H)$ at the point ω and (α, β), the minimality interval of the polynomial P in the set $\mathbf{P}_n(H)$.

If (α, β), (β, γ) is a pair of adjacent minimality intervals associated with the polynomials $P, Q \in \mathbf{P}_n(H)$ then we must have $|P(\beta)| = |Q(\beta)|$, i.e. β is a root of one of the polynomials $P \pm Q$. Hence the endpoints of a minimality interval are algebraic numbers of degree less than or equal to n, with heights not exceeding $2n!H$.

For a fixed ω and $H = 1, 2, \cdots$ we introduce the sequence $P_1, P_2, \cdots$ of all polynomials $P_H \in \mathbf{P}_n(H)$ which are minimal at ω. Clearly,

$$|P_1(\omega)| > |P_2(\omega)| > \ldots > |P_k(\omega)| > \ldots, \tag{3}$$

and there does not exist any polynomial in the set $\mathbf{P}_n = \bigcup_{H=1}^{\infty} \mathbf{P}_n(H)$ which could be inserted in the sequence (3).

For the polynomials (3), Dirichlet's pigeonhole principle yields the inequality

$$|P_k(\omega)| \ll h_{k+1}^{-n}, \tag{4}$$

1) Cf. [55].

where $h_k = h(P_k)$ $(k = 1, 2, \cdots)$.

Let (α_k, β_k) be the minimality interval of P_k which contains ω. If the polynomial P_k does not have a zero in the interval (α_k, β_k) then it must be monotonic in that interval (otherwise P_k' would have a zero in (α_k, β_k); but since $P_k' \in \mathbf{P}_n(h_k)$, this would contradict the fact that (α_k, β_k) is a minimality interval for P_k). Consequently,

$$|P_k(\omega)| > \min(|P_k(\alpha_k)|, |P_k(\beta_k)|) \neq 0. \tag{5}$$

On the other hand, if P_k has a zero at some point κ in (α_k, β_k) then the minimality interval $(\alpha_{k+1}, \beta_{k+1})$ of the polynomial P_{k+1} must have at least one endpoint, say α_{k+1}, within (α_k, β_k). This endpoint α_{k+1} lies between κ and ω (regardless of which of these two numbers is smaller). Therefore

$$|P_k(\omega)| > \min(|P_k(\alpha_1|, |P_k(\beta_1|) \neq 0. \tag{6}$$

and $|P_k(\alpha_{k+1})| = |P_{k+1}(\alpha_{k+1})|$. Hence α_{k+1} (or β_{k+1}) must be a root of either $P_k + P_{k+1}$ or $P_k - P_{k+1}$.

It can be shown easily that (5) and (6) imply, respectively, the relations

$$|P_k(\omega)| \gg h_k^{-2n+1}, \qquad |P_k(\omega)| \gg h_k (h_k h_{k+1})^{-n}.$$

Thus, combining this with (4), we obtain

$$h_{k+1}^{-n} \gg |P_k(\omega)| \gg h_k (h_k h_{k+1})^{-n}. \tag{7}$$

Now we let

$$\sigma_n(\omega) = \overline{\lim_{k \to \infty}} \frac{\ln h_{k+1}}{\ln h_k}.$$

Then it is obvious that always $1 \leq \sigma_n(\omega) \leq \infty$, and thus we have, by (7),

$$n\sigma_n(\omega) + n - 1 \geqslant \overline{\lim_{k \to \infty}} \frac{\ln \dfrac{1}{|P_k(\omega)|}}{\ln h_k} \geqslant n\sigma_n(\omega). \tag{8}$$

The minimality condition for P_k implies that, for all $H < h_{k+1}$,

$$|P_k(\omega)| = \min_{P \in \mathbf{P}_n(H)} |P(\omega)| = w_n(\omega, H),$$

and hence

$$w_n(\omega) = \overline{\lim_{k \to \infty}} \frac{\ln \dfrac{1}{|P_k(\omega)|}}{\ln h_k}.$$

The last equation becomes obvious when we observe that the inequality $|P(\omega)| < h_P^{-w}$ has infinitely many solutions $P \in \mathbf{P}_n$ if and only if it has infinitely many solutions from the sequence (3).

Therefore we may conclude from (8) that

$$\sigma_n(\omega) \leqslant \Theta_n(\omega) \leqslant \sigma_n(\omega) + 1 - \frac{1}{n} \quad (n = 1, 2, \ldots). \tag{9}$$

This relation implies, in particular, that the two quantities $\sigma_n(\omega)$ and $\Theta_n(\omega)$ are either both finite or both infinite and that Mahler's classification can be based on the quantity $\sigma_n(\omega)$ rather than $\Theta_n(\omega)$.

For $n = 1$ the relation (9) is well known in connection with the theory of continued fractions (cf. [21], [6]).

We note that the equations

$$\sigma_n(\omega) = \Theta_n(\omega) = 1 \quad (n = 1, 2, \ldots)$$

follow from (4) for almost all numbers ω.

An inequality analogous to (9) may be obtained by considering polynomials of bounded heights and increasing degrees. Again a sequence of "minimal" polynomials $P_1, P_2, \cdots$ may be defined. If this sequence P_k has the degree n_k then a classification like the one we discussed above may be based on the quantity

$$\tau(H; \omega) = \overline{\lim_{k \to \infty}} \frac{\ln n_{k+1}}{\ln n_k}$$

B. FIELDS OF p-ADIC NUMBERS

On the number of cubic polynomials with bounded height and a given discriminant. [1] The cubic case of Mahler's conjecture may be treated by a strictly arithmetical method based on the inequalities (10) and (12) below. Let $\mathbf{P}_3(h)$ be the set of polynomials $P(x) = a_0 + a_1x + a_2x^2 + a_3x^3$ satisfying the condition $\max(|a_0|, |a_1|, |a_2|) \leq a_3 = h$. Then for any integer D, the number $V(D)$ of polynomials $P \in \mathbf{P}_3(h)$ with discriminant $D(P) = D$ satisfies

$$V(D) \ll h^{\varepsilon}(h^*, d)\, N\{x^3 = y^2 + A\}, \tag{10}$$

where h^* and d are the greatest integers whose squares are divisors of h and D, respectively, and $N\{\cdots\}$ denotes the number of solutions of the diophantine equation

---

1) Cf. [55].

$$x^3 = y^2 + A, \quad A = 2^4\, 3^3\, h^2 D. \tag{11}$$

The inequality (10) follows immediately from the identity

$$\left[4\,(a_2^2 - 3a_1 h)\right]^3 = \left[4\,(27a_0h^2 - 9a_1a_2h + 2a_2^3)\right]^2 + 2^4\, 3^3\, h^2 D\,(P),$$

which is a direct consequence of a well-known identity due to Cayley (see [10], p. 316 (translation p. 415)).

Concerning estimates on the number of solutions of (11), it is possible to show by a refinement of the Mordell-Siegel method (see [63], [65]) that

$$N\,\{x^3 = y^2 + A\} \ll h^{\varepsilon}\, 3^{\sigma(-g)}$$

where g is the square-free part (the product of all prime divisors which occur in the first power only) of the number $3D$; furthermore $3^{\sigma(-g)}$ is the number of classes X in the ideal-class group Γ of the field $Q(\sqrt{-g})$ satisfying the equation $X^3 = E$ (E denotes the principal class, Q the rational number field; $\sigma(-g)$ may be defined as the basis number of order 3 in the group Γ, i.e. as the number of basis elements of the class group whose order is divisible by 3). Since the elements of the group Γ which satisfy the condition $X^3 = E$ form a subgroup of order $3^{\sigma(-g)}$, it follows that $3^{\sigma(-g)}$ cannot be greater than the number of ideal classes of the field $Q(\sqrt{-g})$. This number is $\ll g^{1/2+\epsilon}$ (see, for example, the closing remark in the paper [1]), and hence we have in all cases

$$N\,\{x^3 = y^2 + A\} \ll h^{\varepsilon}\, |g|^{\frac{1}{2}+\varepsilon} \tag{12}$$

C. FIELDS OF POWER SERIES

Linear diphantine approximations and the theory of matrices. Let K be a field, x a transcendental element over K, $R = K[x]$ the ring of polynomials in X over K, $K\langle x\rangle$ the field of formal power series of the form

$$\omega(x) = \sum_{s=l}^{\infty} a_s x^{-s}, \quad a_s \in K \quad (s = l,\ l+1,\ \ldots); \tag{13}$$

furthermore, let $q > 1$ be an arbitrary, but henceforth fixed number, and let

$$|\omega(x)| = \begin{cases} 0, & \text{if in (13) } a_s = 0 \quad (s = l,\ l+1,\ \ldots), \\ q^{-l}, & \text{if in (13) } a_l \neq 0. \end{cases}$$

We define

$$[\omega(x)] = \sum_{s\leqslant 0} a_s x^{-s}, \quad \{\omega(x)\} = \sum_{s>0} a_s x^{-s}$$

$$(\omega(x) = [\omega(x)] + \{\omega(x)\}).$$

(thus $\omega(x) = [\omega(x)] + \{\omega(x)\}$), calling $[\omega(x)]$ and $\{\omega(x)\}$, respectively, the integral and the fractional part of $\omega(x)$.

Then we consider the linear inequality

$$|\{L(a_1, a_2, \ldots, a_n)\}| < q^{-m}, \quad \max_{(i)} |a_i| \leq q^h \quad (i = 1, 2, \ldots, n), \tag{14}$$

where

$$L(a_1, a_2, \ldots, a_n) = a_1 \omega_1 + a_2 \omega_2 + \ldots + a_n \omega_n,$$
$$a_i \in R, \quad \omega_i \in K\langle x \rangle, \quad \max_{(i)} |\omega_i| \leq 1 \quad (i = 1, 2, \ldots, n).$$

Now let

$$\omega_i(x) = \sum_{k=0}^{\infty} \beta_{ik} x^{-k},$$
$$a_i(x) = \alpha_{i0} + \alpha_{i1} x + \ldots + \alpha_{ih} x^h \quad (i = 1, 2, \ldots, n).$$

Then we have

$$L(a_1, a_2, \ldots, a_n) = \sum_{\nu=-h}^{\infty} \lambda_\nu x^{-\nu}, \quad \lambda_\nu = \sum_{i=1}^{n} S_{i\nu},$$
$$S_{i\nu} = \sum_{k=\nu_0}^{h} \alpha_{ik} \beta_{ik+\nu}, \quad \nu_0 = \max(0, -\nu). \tag{15}$$

It is evident that the greatest number m for which the inequality (14) is solvable with a given h, is identical with the greatest number m for which the system of linear homogeneous equations (in the α_{ik})

$$\lambda_0 = 0, \quad \lambda_1 = 0, \quad \ldots, \quad \lambda_m = 0 \tag{16}$$

has a nontrivial solution. This assertion follows directly from the relations (14) and (15).

The matrix $M_n(m, h)$ of the system (16) is composed of Hankel matrices

$$H_{mh}^{(i)} = \begin{bmatrix} \beta_{i1} & \beta_{i2} & \cdots & \beta_{i1+h} \\ \beta_{i2} & \beta_{i3} & \cdots & \beta_{i2+h} \\ \cdots & \cdots & \cdots & \cdots \\ \beta_{im} & \beta_{im+1} & \cdots & \beta_{im+h} \end{bmatrix}$$

and has the form

$$M_n(m, h) = \left[H_{mh}^{(1)} H_{mh}^{(2)} \dots H_{mh}^{(n)}\right].$$

Let $r_n(m, h)$ be the rank of the matrix $M_n(m, h)$. Then the system (16), and hence also the inequality (15), has a nontrivial solution if and only if

$$r_n(m, h) < n(h+1). \tag{17}$$

For a given h we define the greatest $m = m(h)$ for which (17) holds. The nondecreasing function $m(h)$ describes the behavior of the minima of the linear form $L(a_1, a_2, \cdots, a_n)$, i.e. for any positive integer h, the value $m(h)$ is the greatest number for which the system (14) has a solution. Obviously we may choose $m = n(h+1) - 1$.

Now we consider the system

$$\max(|\{\omega_1 g\}|, |\{\omega_2 g\}|, \dots, |\{\omega_n g\}|) < q^{-m}, \ |g| \leqslant q^h, \tag{18}$$

where $g = g(x) \in R$. By an argument like the one given above we find that the system (18) has a nontrivial solution if and only if the rank of the matrix

$$N_n(m, h) = \begin{bmatrix} H_{mh}^{(1)} \\ H_{mh}^{(2)} \\ \cdots \\ H_{mh}^{(n)} \end{bmatrix}$$

does not exceed h. Clearly, we may choose $m = [h/n]$.

From the obvious relation $(H_{mh}^{(i)})' = H_{h+1, m-1}^{(i)}$, where the prime denotes the transpose, we obtain the relations

$$M_n'(m, h) = N_n(h+1, m-1),$$

$$N_n'(m, h) = M_n(h+1, m-1).$$

These equations show that the solvability of the system (14) follows immediately from the solvability of the system (18) and vice versa.

In the case of real numbers such a connection is established by Hinčin's "principle of transfer".

The principle of transfer. Suppose K is a finite field with q elements and that the number $q > 1$ satisfies the inequality

$$\max(|\{\omega_1 g\}|, \dots, |\{\omega_n g\}|) < q^{-m}, \ |g| = q^h, \tag{19}$$

in which we may, in the light of a remark made above, assume that

$$m \geqslant m_0 = \left[\frac{h}{n}\right]. \tag{20}$$

The inequality (19) may be rewritten in the form

$$|\omega_1 g - a_1| < q^{-m}, \ \ldots, \ |\omega_n g - a_n| < q^{-m}, \quad |g| = q^h, \tag{21}$$

where $a_i = [\omega_i g] \in R$ $(i = 1, 2, \cdots, n)$. Multiplying the inequalities (21) by polynomials $c_i \in R$ for which

$$|c_i| \leqslant q^{m_0} \quad (i = 1, 2, \ldots, n), \tag{22}$$

we obtain the relation

$$|g(\omega_1 c_1 + \ldots + \omega_n c_n) - L(c_1, \ldots, c_n)| < \max_{(i)} |c_i| \, q^{-m} \leqslant q^{m_0 - m}, \tag{23}$$

where $L(c_1, c_2, \cdots, c_n) = a_1 c_1 + a_2 c_2 + \cdots + a_n c_n$. Because of (22) every c_i assumes at most q^{m_0+1} different values; hence the form $L(c_1, \cdots, c_n)$ assumes not more than $q^{n(m_0+1)}$ values. Since the number of residue classes mod g is equal to $|g| = q^h$ and since, by (20), $n(m_0 + 1) \geq n(h/n - 1) + n = h$, there exists a pair of different vectors $\bar{c} = (c_1, \cdots, c_n)$ for which the values of $L(c_1, \cdots, c_n)$ are congruent mod g. Consequently, we have from (23)

$$|\omega_1 c_1' + \ldots + \omega_n c_n' - c| < q^{-h+m_0-m}$$

for some c and $c_i' \in R$ with $\max_{i=1,n} |c_i'| \leq q^{m_0}$. Thus, if the system (19) has a solution, so does the system

$$\|\{\omega_1 c_1 + \ldots + \omega_n c_n\}\| < q^{-h+m_0-m},$$
$$\max_{(i)} |c_i| \leqslant q^{m_0} \quad (i = 1, 2, \ldots, n).$$

Hence we may conclude that if for a given $\epsilon > 0$ there are infinitely many solutions of the inequality

$$\max(\|\{\omega_1 g\}\|, \ \ldots, \ \|\{\omega_n g\}\|) < q^{-\frac{h}{n}(1+\varepsilon)}, \quad |g| = q^h,$$

then there also exist infinitely many solutions of the inequality

$$\|\{\omega_1 c_1 + \ldots + \omega_n c_n\}\| < H^{-n\left(1 + \frac{\varepsilon}{n}\right)}$$

with

$$\max_{i=1,\cdots,n} |c_i| \le qH, \quad H = q^{h/n}.$$

It should be noted that the argument just outlined also applies in the cases of real and of p-adic numbers.

Algebraic approximations. If κ is of finite degree over $R = K[x]$ then the height $h(\kappa)$ may be defined as the height of the minimal polynomial of κ, i.e. the (unique) height of a primitive polynomial of lowest degree with κ as a root. For transcendental elements $\omega \in K\langle x\rangle$ we define $w_n(\omega)$ as the supremum of the set of those $w > 0$ for which there exist infinitely many solutions of the inequality $|\omega - \kappa| < h_\kappa^{-w}$ by algebraic numbers κ with degrees not exceeding n. It can be shown (see [62], [55]) that the quantities $w_n(\omega)$ and $w_n^*(\omega)$ are related to each other by several inequalities. For example,

$$w_n(\omega) \gg w_n^*(\omega),$$

$$w_n^*(\omega) \gg \frac{w_n(\omega)}{w_n(\omega) - n + 1}$$

and if K has characteristic zero, then also

$$w_n^*(\omega) \gg w_n(\omega) - n + 1,$$

$$w_n^*(\omega) \gg \frac{1}{2}(w_n(\omega) + 1) \quad (n = 1, 2, \ldots).$$

In particular, if K is the field of complex numbers, then the field $K\langle x\rangle$ contains those functions $f(x)$ which are regular in a neighborhood of infinity. A valuation on those functions may be defined as q^α, where q is an arbitrary real number greater than 1 and

$$\alpha = \overline{\lim_{R\to\infty}} \frac{\ln \frac{1}{|f(R)|}}{\ln R}.$$

Consequently, the inequalities mentioned above express statements on the approximation of functions which are regular in a neighborhood of infinity by algebraic functions of a given degree. It is possible to classify the power series in the manner of Mahler and Koksma, and this leads to a classification of analytic functions in terms of their behavior with respect to approximation by algebraic functions.

CONCLUSION

§1. SOME COROLLARIES TO THE RESULTS OBTAINED

We wish to mention some consequences which follow from the validity of Mahler's and Kasch's conjectures, i.e. from the equations

$$w_n(\omega) = n \quad (n = 1, 2, \ldots) \tag{1}$$

for almost all real numbers and

$$w_n(\omega) = \frac{n-1}{2} \quad (n = 2, 3, \ldots) \tag{2}$$

for almost all complex numbers.

Corollary 1. (Wirsing's conjecture [80]). *Let* $w_n^*(\omega)$ *be defined as in the Introduction (cf. §1). Then*

$$w_n^*(\omega) = n \quad (n = 1, 2, \ldots) \tag{3}$$

for almost all real numbers, and

$$w_n^*(\omega) = \frac{n-1}{2} \quad (n = 2, 3, \ldots) \tag{4}$$

for almost all complex numbers.

Proof. From (1) and (2) we have, respectively, by the inequalities (5) and (6) of the Introduction, $w_n^*(\omega) \geq n$ for all real and $w_n^*(\omega) \geq (n-1)/2$ for all complex transcendental numbers. But since $w_n(\omega) \geq w_n^*(\omega)$, strict inequality in the last two relations is only possible for sets of measure zero, and hence (3) and (4) follow.

Thus, given any $\epsilon > 0$, the inequality $|\omega - \kappa| < h_\kappa^{-n-1+\epsilon}$ is for almost all real ω satisfied by infinitely many algebraic numbers κ with degrees not exceeding n. In the case of complex numbers the same is true for the inequality

$$|\omega - \kappa| < h_\kappa^{-\frac{n+1}{2}+\varepsilon}$$

Corollary 2. *For any real* α *let* $\|\alpha\|$ *denote the distance between* α *and the nearest integer. Let* $\lambda_n(\omega)$ *be the supremum of the set of those* $\lambda > 0$ *for which the system*

$$\max(\|\omega q\|, \|\omega^2 q\|, \ldots, \|\omega^n q\|) < q^{-\lambda} \tag{5}$$

of inequalities is satisfied by infinitely many natural numbers q. *Then*

$$\lambda_n(\omega) = \frac{1}{n} \quad (n = 1, 2, \ldots)$$

for almost all numbers.

Proof. This result is obtained from (1) by applying Hinčin's "principle of transfer" (see [18], [45]). It was well known (see [8]) that this assertion is equivalent to Mahler's conjecture for real numbers before the latter was proved.

Corollary 3 (see [68]). *For any real* ω *let* $\mu_n(\omega)$ *denote the supremum of the set of those* $\mu > 0$ *for which the system*

$$\max(|P(\omega)|, |P'(\omega)|, \ldots, |P^{(n-1)}(\omega)|) < h_P^{-\mu} \tag{6}$$

of inequalities (where $P^{(i)}(x)$ *denotes the ith derivative of* $P(x)$*) is satisfied by infinitely many polynomials* $P(x)$*, with integer coefficients, of degrees not exceeding* n. *Then the equations*

$$P^{(0)}(x) = P(x), \quad P^{(i)}(x) = \frac{d}{dx} P^{(i-1)}(x) \quad (i = 1, 2, \ldots, n-1)$$

hold for almost all numbers.

Proof. In view of Corollary 2 it suffices to show that $\mu_n(\omega) = \lambda_n(\omega)$ $(n = 1, 2, \cdots)$ for any ω.

In the conditions (5) we let

$$\|\omega^i q\| = |\omega^i q - a_i| \quad (i = 1, 2, \ldots, n) \tag{7}$$

and we define the polynomials

$$P(x) = qx^n - a_1 \binom{n}{1} x^{n-1} + a_2 \binom{n}{2} x^{n-2} - \ldots + (-1)^n a_n. \tag{8}$$

Then for $i = 1, 2, \cdots, n-1$ we have the identities

$$P^{(i)}(x) = q \binom{n}{i} i!\, x^{n-i} + \sum_{k=1}^{n-i} (-1)^k \binom{n}{k} \binom{n-k}{i} i!\, a_k x^{n-k-i} =$$

$$= n(n-1)\dots(n-i+1)\left(qx^{n-i} + \sum_{k=1}^{n-1}(-1)^k\binom{n-i}{k}a_k x^{n-k-i}\right).$$

By substituting $x = \omega$ in the last formula and eliminating the a_k by means of (7) we obtain from (5) an identity of the type

$$\frac{P^{(i)}(\omega)}{n(n-1)\dots(n-i+1)} = q\omega^{n-i}\left(1+\sum_{k=1}^{n-1}(-1)^k\binom{n-i}{k}\right)$$

$$+ c(n, \omega)q^{-\lambda} = q\omega^{n-i}(1-1)^{n-i} + c(n, \omega)q^{-\lambda} = c(n, \omega)q^{-\lambda}.$$

Thus (5) implies

$$\max(|P(\omega)|, |P'(\omega)|, \dots, |P^{(n-1)}(\omega)|) \ll q^{-\lambda}.$$

and thus $\mu_n(\omega) \geq \lambda_n(\omega)$.

The converse inequality is established by the following argument. The system (6) has the form

$$|a_0 + a_1\omega + a_2\omega^2 + \dots + a_n\omega^n| < h^{-\mu},$$

$$|a_1 + 2a_2\omega + \dots + na_n\omega^{n-1}| < h^{-\mu},$$

$$\dots\dots\dots\dots\dots\dots\dots\dots$$

$$(n-1)!\,|a_{n-1} + na_n\omega| < h^{-\mu}.$$

By elementary operations it can be transformed into the form

$$|a_0 + c_1 a_n\omega^n| < c(n, \omega)h^{-\mu}$$

$$|a_1 + c_2 a_n\omega^{n-1}| < c(n, \omega)h^{-\mu}$$

$$\dots\dots\dots\dots\dots\dots$$

$$|a_{n-1} + c_n a_n\omega| < c(n, \omega)h^{-\mu},$$

where $c_1, c_2, \cdots, c_n$ are integers which depend only on n. Now we let $c = c_1 c_2 \cdots c_n$, $s = c/c_i$, and we multiply the first inequality by s_1, the second by s_2, etc. This leads to the system

$$\max(\|\omega ca_n\|, \|\omega^2 ca_n\|, \dots, \|\omega^n ca_n\|) \ll (ca_n)^{-\mu},$$

which is essentially the same as (5).

Still further corollaries are possible, but we will not discuss them here.

§2. GENERAL CONSEQUENCES

First we wish to remark that the following modifications of Mahler's problem for complex numbers is possible. Let $K = Q(\sqrt{-d})$ be an imaginary quadratic number-field and let C be the ring of integers in K. Then for every complex transcendental number ω we define the quantity $w_n(\omega)$ as the supremum of the set of those $w > 0$ for which there exist infinitely many solutions of the inequality

$$|a_0 + a_1\omega + \dots + a_n\omega^n| < h^{-w}, \quad h = \max(|a_0|, \dots, |a_n|) \tag{9}$$

by numbers $a_i \in C$ $(i = 0, 1, \cdots, n)$. It is easy to show by means of Dirichlet's pigeonhole principle that always

$$w_n(\omega) \geqslant n \quad (n = 1, 2, \dots). \tag{10}$$

This gives rise to the following problem: in the inequality (10) does the equality sign hold for almost all ω? This is in fact true and can be shown by a minor modification of the proof given in Part I, but with the condition that the number of ideal-classes in K be equal to 1. This condition is important because our method of proof requires the transition in (9) to the case where only such polynomials are allowed as solutions which are irreducible over C. We accomplish this transition by an application of Gauss's Lemma (see, for example, [81] or [33]). It is not impossible that the analog of Mahler's conjecture described above might remain valid in the case where the class number of the field K is greater than 1.

Turning now to more abstract considerations, let us assume that C is a commutative ring with identity and without zero divisors, i.e. a domain of integrity. Let K_0 be the quotient field of C, and let K be any extension field of K_0. Thus $C \subseteq K_0 \subseteq K$. We are interested in knowing which conditions the ring C must satisfy in order that it can serve as a basis for a natural approach to studying "diophantine" approximations of the elements of the field K by elements of the ring C. It is clear that some metric has to be defined in K and consequently also in C such that the notion of "distance" can be defined in K with respect to this metric. Suppose that a valuation $|\cdots|$ on the field is given. Since the ring C has to play in K a role analogous to the role played by the ring of integers in the field of real numbers, it is natural to require that the valuation be discrete on C, i.e. for any $x > 0$, the number $N(x)$ of elements $\omega \in C$ with $|\omega| \leq x$ has to be finite.

The following assertion is implied by a theorem of Ostrowski [49] (see [78], Chapter X, §75): If on a commutative ring C of characteristic zero, with identity and without zero divisors there exists a discrete valuation (in the sense just defined), then C is isomorphic either to the ring of rational integers or to a subring of the ring of integers in some imaginary quadratic field $Q(\sqrt{-d})$.

Indeed, C contains the ring Z of rational integers. If $t \in C$ then $|t| \geq 1$, since in the case $|t| < 1$ the sequence of the elements $t^k \in C\,(k = 0, 1, \cdots)$ would contradict the discreteness of the valuation. Now we distinguish two cases:

1) If $t \in C$ is an element of degree 1 over Z then $t \in Z$.

Proof. Indeed, t is of the form m/n, $(m, n) = 1$. We determine integers x and y such that $mx + ny = 1$. Then $1/n = mx/n + y \in C$, $|1/n| \geq 1$; hence $n = \pm 1$.

2) If $t \in C$, $t \notin Z$, then t has degree 2 over Z.

Proof. It is clear that the field Q of rational numbers is contained in K_0. Since $t \in C$, we have $Q(t) \subseteq K_0$. The field K_0 and thus also the field $Q(t)$ are archimedean. By Ostrowski's theorem there is a continuous isomorphism between some subfield of the complex number field under the valuation by absolute values. Let τ be the image of t under this isomorphism. (Thus τ is a real or complex number.) Trivially, t and τ have the same degree over Z. If the degree of τ over Z is greater than 2 then there exists an infinite sequence of polynomials P_ν of degree 2 over Z, with rational integer coefficients such that $\lim_{\nu\to\infty} P_\nu(\tau) = 0$ and such that the numbers $P_\nu(\tau)$ are different from each other (if τ is a real number then we can choose polynomials of the first degree over Z). Because of the continuity of the isomorphism we have $\lim_{\nu\to\infty} P_\nu(t) = 0$; furthermore, $P_\nu(t) \in C$ and all the numbers $P_\nu(t)$ are different. Thus the valuation cannot be discrete. Consequently, the degree of t over Z is not greater than 2, and since it is not equal to 1, it must be 2.

This reasoning shows that τ must necessarily be a nonreal complex number. Indeed, otherwise we could select polynomials P_ν of degree 1 over Z such that $P_\nu(\tau) \to 0$, and again we would obtain a contradiction to the discreteness of the valuation.

If now $t_1 \in C$, $t_1 \notin Z$, then t_1 must belong to $Q(t)$ since otherwise $Q(t, t_1)$ would be of degree 4 over Q and thus there would exist in $Q(t, t_1)$ some element of degree 4. Consequently, there would exist in C an element of degree 4 over Z, but this is impossible in case 2). Hence $C \subseteq Q(t)$ and therefore $K_0 = Q(t)$.

Now we observe that τ is an integer of the field K_0. Indeed, let $a_0 + a_1\tau + a_2\tau^2 = 0$. By the isomorphism $t \leftrightarrow \tau$ we have the relations

$$C \ni t_1 + t_2 \leftrightarrow \tau_1 + \tau_2 = -\frac{a_1}{a_2},$$

$$C \ni t_1 t_2 \leftrightarrow \tau_1\tau_2 = \frac{a_0}{a_2};$$

'urthermore, the elements of the rational number field Q are left invariant by this isomorphism. Thus $a_1/a_2 \in C$, $a_0/a \in C$, and by the assertion proved under 1) it follows that $a_2 | a_1$, $a_2 | a_0$. Hence τ is an integer of K_0.

Consequently, t is integral over Z and the same is true for all $t_1 \in C$. Thus $C \subseteq Z(t)$ and this completes the proof of our assertion.

If the characteristic of the ring C is different from zero then the structure of the ring C is more complicated than in the case of characteristic zero. In any case it is obvious that the valuation on C is non-archimedean and that C contains the subring $K^*[x]$ of polynomials over some finite field K^*.

The case of fields with a non-archimedean valuation was treated in Part II under sufficiently general assumptions (we only assumed the ground field to be locally compact).

In making these final observations we wish to remark that statements along the lines of Mahler's conjecture reflect a general property of different kinds of algebraic fields. In fact the investigation of Part I can be carried over to the case of approximations by integers in an imaginary quadratic field (at least if it has class-number 1), and, by the remarks made above, the same problem can be treated exhaustively in the general case of domains with characteristic zero. The investigation of Part II develops sufficiently far the case of fields with a non-archimedean valuation, i.e. domains with prime characteristic.

§3. NEW PROBLEMS AND CONJECTURES

Let $\|\omega\|$ denote the distance between the real number ω and the nearest integer, let $\psi(q)$ denote a positive function of the natural numbers (for example $\psi(q) = q^{-\lambda}$, $\lambda > 0$) and consider the system

$$\max\{\|\Omega_1 q\|, \|\Omega_2 q\|, \ldots, \|\Omega_n q\|\} < \psi(q) \qquad (11)$$

of diophantine inequalities. In the case of mutually independent numbers Ω_1, $\Omega_2, \cdots, \Omega_n$ a metrical theory of the system (11) has been developed in great detail by many authors (cf. Hinčin [19], Duffin and Schaeffer [11], Cassels [6], [4], Schmidt [51]).

Undoubtedly the case of dependent numbers $\Omega_1, \Omega_2, \cdots, \Omega_n$ deserves interest also (cf. [57], [53]). Precisely speaking, we have in mind a system $\Omega_1, \Omega_2, \cdots, \Omega_n$ of real functions, defined on the product space $R^k = R \times R \times \cdots \times R$ (R is the space of real numbers ω) with $1 \leq k < n$. We assume that the functions

$$\Omega_i = \Omega_i(\omega_1, \omega_2, \dots, \omega_k) \quad (i = 1, 2, \dots, n)$$

are measurable on R^k and linearly independent over the field of rational numbers.

Hinčin's "principle of transfer" permits us, whenever convenient, to switch from the system (11) to an inequality of the form

$$\begin{gathered} \|a_1 \Omega_1 + a_2 \Omega_2 + \dots + a_n \Omega_n\| < h^{-v}, \\ h = \max(|a_1|, \dots, |a_n|). \end{gathered} \tag{12}$$

In particular, if $\Omega_1 = \omega$, $\Omega_2 = \omega^2, \cdots, \Omega_n = \omega^n$, then we arrive at the case of Mahler's conjecture which we have dealt with.

Let $v(\Omega_1, \Omega_2, \cdots, \Omega_n)$ denote the supremum of the set of those numbers $v > 0$ for which the inequality (12) has infinitely many solutions. Clearly, Dirichlet's pigeonhole principle leads to the inequality $v(\Omega_1, \cdots, \Omega_n) \geq n$ for all vectors $\bar{\omega} = (\omega_1, \omega_2, \cdots, \omega_n)$ for which $\Omega_1, \cdots, \Omega_n$ are linearly independent over the field of rational numbers. The question arises naturally whether the equation

$$v(\Omega_1, \dots, \Omega_n) = n. \tag{13}$$

holds for almost all $\bar{\omega} \in R^k$. The answer to this question is undoubtedly connected with intrinsic properties of the functions $\Omega_1, \Omega_2, \cdots, \Omega_n$ and in many cases it appears to be very difficult. It would be of significant interest to prove the finiteness of the quantity $v(\Omega_1, \cdots, \Omega_n)$ for almost all $\bar{\omega} \in R^k$ or to establish an inequality of the form

$$v(\Omega_1, \dots, \Omega_n) < c(n) \tag{14}$$

for almost all $\bar{\omega} \in R^k$.

Generally speaking, the problem under consideration is the more complicated the smaller the ratio k/n is. If $k(n) \leq k < n$, where the function $k(n)$ does not increase too slowly for increasing n (for example, $k(n) >> n^{\delta}$, $\delta > 0$) then in many cases the equation (13) may be shown to be true for almost all $\bar{\omega} \in R^k$ by means of the method of trigonometric sums, using results of I. M. Vinogradov [71] and other mathematicians (cf. [22]). In this connection, any improvement of the estimates for trigonometric sums leads to a lowering in the order of growth of the permissible functions $k(n)$.

In particular, in the case where the functions $\Omega_1, \cdots, \Omega_n$ are polynomials in $\omega_1, \cdots, \omega_k$, a reduction of the power (to an absolute constant) in the estimates pertaining to trigonometric sums with polynomial exponents would lead to a function $k(n)$ bounded (from below) by an absolute constant. The trigonometric sums in question have the form

$$S = \sum_{a_1, \ldots, a_n} \exp 2\pi iqF\left(\frac{a_1}{q}, \ldots, \frac{a_n}{q}\right), \tag{15}$$

where q is a natural number; furthermore,

$$F(x_1, \ldots, x_n) = c_1\Omega_1(x_1, \ldots, x_n) + \ldots + c_n\Omega_n(x_1, \ldots, x_n),$$

with integers c_i satisfying

$$\max_{i=1, \cdots, n} |c_i| \ll q^{\frac{1}{n}},$$

and with the summation (15) extended over all integers a_i with

$$\max_{i=1, \cdots, n} |a_i| \ll q.$$

Concerning the reduction of the problem (11) to estimates of the summation (15) the reader is referred to the author's paper [57], where systems of the form

$$\max_{i,j=1, \cdots, n} (\|\omega_i q\|, \|\omega_i \omega_j q\|) < q^{-\lambda}$$

are investigated (also see "An application", §2).

The first application of the method of trigonometric sums (in the form due to I. M. Vinogradov) to the solution of problems of the form (11) was made by J. P. Kubilius (cf. [31], [32]).

On the other hand, it is likely that the method used in this book for solving Mahler's problem will lend itself to the investigation of many problems of the form (12) in the case where $k(n)$ is relatively small in relation to n. Considering the inequality (12) with a suitable v (for example, $v = n - 1 - \epsilon$) and introducing "essential" and "inessential" domains in some part of the space R^k, we can again, on the basis of an appropriate "lemma on partial coverings", exclude the inessential domains from our consideration, and for the investigation of the essential domains one has to determine the order of magnitude of the volume of the domains defined by the inequalities

$$\|a_1 \Omega_1 + \ldots + a_n \Omega_n\| < h^{-v}, \quad \max_{i=1,\cdots,n} |\omega_i| \ll 1.$$

In addition to this procedure it will probably be necessary, for some problems, to treat special cases of the last inequality separately (as, for example, classes of the second kind in Mahler's problem).

We will formulate some questions which are directly related to Mahler's problem and whose solution would, in our opinion, be of the foremost interest at present.

Problem A. *Let $m_1, m_2, \cdots, m_n$ be mutually different natural numbers, let ω be a transcendental number and let*

$$v_n(\omega) = v(m_1, m_2, \ldots, m_n; \omega)$$

be the function $v(\Omega_1, \cdots, \Omega_n)$ defined above, with the parameters

$$\Omega_1 = \omega^{m_1}, \ \Omega_2 = \omega^{m_2}, \ \ldots, \ \Omega_n = \omega^{m_n}.$$

Does the equation

$$v_n(\omega) = n \quad (n = 1, 2, \ldots) \tag{16}$$

hold for almost all real ω, regardless of the choice of the numbers $m_1, \cdots, m_n$?

Problem B.1. *Let $n_1, n_2, \cdots, n_k$ be arbitrary natural numbers and let $v(\omega_1, \omega_2, \cdots, \omega_k)$ be the function $v(\Omega_1, \cdots, \Omega_n)$ defined above, with*

$$\Omega_j = \omega_1^{i_1} \omega_2^{i_2} \ldots \omega_n^{i_n} \quad (j = 1, 2, \ldots, n),$$

where $i_1, i_2, \cdots, i_n$ satisfy the conditions

$$0 \leqslant i_1 \leqslant n_1, \quad 0 \leqslant i_2 \leqslant n_2, \ \ldots, \ 0 \leqslant i_k \leqslant n_k,$$

$$i_1 + i_2 + \ldots + i_k \neq 0, \quad n = (n_1 + 1)(n_2 + 1) \ldots (n_k + 1) - 1.$$

Does the equation

$$v(\omega_1, \ldots, \omega_k) = n$$

hold for almost all $\bar{\omega} = (\omega_1, \omega_2, \cdots, \omega_k)$, *regardless of the choice of the numbers* $n_1, n_2, \cdots, n_k$?

Problem B. 2. *Let* m *be an arbitrary natural number and let* $v(\omega_1, \cdots, \omega_k)$ *be the function* $v(\Omega_1, \cdots, \Omega_n)$ *defined above, with*

$$\Omega_j = \omega_1^{i_1}\omega_2^{i_2} \ldots \omega_n^{i_n} \quad (j = 1, 2, \ldots, n),$$

where $i_1, i_2, \cdots, i_n$ *satisfy the conditions*

$$0 \neq i_1 + i_2 + \ldots + i_k \leqslant m,$$
$$i_1 \geqslant 0, \ldots, i_k \geqslant 0,$$
$$n = \binom{m+k}{k} - 1.$$

Does the equation

$$v(\omega_1, \ldots, \omega_k) = n \tag{17}$$

hold for almost all $\bar{\omega} = (\omega_1, \omega_2, \cdots, \omega_k)$?

The answers to these questions are affirmative in some special cases. For example, we know that (16) is valid for $m_1 = 1, m_2 = 2, \cdots, m_n = n$ and that (17) holds for $m = 1, 2$ with an arbitrary $k = 1, 2, \cdots$ (cf. [57]).

W. Schmidt [51] obtained an interesting result in the case $n = 2$, $k = 1$: If the curvature of the curve $\Gamma = (\Omega_1(\omega), \Omega_2(\omega))$ is almost everywhere different from zero, then (13) is true almost everywhere.

Recently Baker [82], following the general scheme of the author's work ([65]–[68]) but introducing a number of ingenious simplifications and sharpenings, proved that, given any natural number n and a monotonically decreasing function $\psi(h)$ for which the series $\sum_{h=1}^{\infty} \psi(h)$ converges, the inequality

$$|P(\omega)| < \psi^n(h), \quad h = h(P) \tag{18}$$

is, for almost all real numbers ω, satisfied by at most finitely many polynomials P of degrees not exceeding n, with integer coefficients. The same is true in the case of complex numbers (with $n = 2, 3, \cdots$) if the exponent n in (18) is replaced by $(n-1)/2$. Probably a similar result is true for locally compact fields with a non-archimedean valuation.

AN APPLICATION

§1. SIMULTANEOUS LINEAR DIOPHANTINE APPROXIMATIONS

Let R^n be the n-dimensional euclidean space, let E_n be the unit cube in R^n and let Ω_{mn} be the space of $(m \times n)$-matrices over E_1, i.e. the set of matrices

$$\omega = \begin{pmatrix} \omega_{11}\omega_{12} & \cdots & \omega_{1n} \\ . & . & . & . & . & . & . & . \\ \omega_{m1}\omega_{m2} & \cdots & \omega_{mn} \end{pmatrix} \tag{1}$$

with $0 \leq \omega_{ij} < 1$ $(i = 1, 2, \cdots, m;\ j = 1, 2, \cdots, n)$. On the space Ω_{mn} Lebesgue measure may be defined by means of the natural one-to-one mapping of Ω_{mn} onto the mn-dimensional unit cube, i.e. by

$$\omega \leftrightarrow (\omega_{11},\ \omega_{12},\ \cdots\ \omega_{1n},\ \ldots,\ \omega_{m1},\ \omega_{m2},\ \ldots,\ \omega_{mn}).$$

Let the class of measurable sets in Ω_{mn} be denoted by $\mathbf{A}_{mn}$.

With each vector $\mathbf{a} \in R^n$, $\mathbf{a} \neq (0)$, we associate the following mapping $T = T(\mathbf{a})$ of the space Ω_{mn} onto Ω_{1m}:

$$\omega \to (\mathbf{a} \cdot \bar{\omega}_1,\ \ldots,\ \mathbf{a} \cdot \bar{\omega}_n) \bmod 1, \tag{2}$$

where the $\bar{\omega}_i$ are the row vectors of the matrix (1), and $\mathbf{a} \cdot \bar{\omega}_i$ denotes the scalar product of the vector $\mathbf{a}$ with $\bar{\omega}_i$ $(i = 1, 2, \cdots, n)$.

If $A \in \mathbf{A}_{1m}$, then let $T^{-1}A$ denote the inverse image of A in Ω_{mn} with respect to the mapping (2). Later on we will show that T is measure preserving, i.e.,

$$|T^{-1}A| = |A|, \tag{3}$$

and furthermore, if $\mathbf{a}_1$, $\mathbf{a}_2$ are two linearly independent vectors in R^n, that the sets $T^{-1}A$ are stochastically independent in the following sense: If $A_1, A_2 \in \mathbf{A}_{1m}$, $T_1 = T(\mathbf{a}_1)$, $T_2 = T(\mathbf{a}_2)$, then

$$|T_1^{-1}A_1 \cap T_2^{-1}A_2| = |A_1|\,|A_2| \tag{4}$$

(here and in the sequel the measure of measurable set A is denoted by $|A|$).

The equations (3) and (4) allow us to prove the following theorems.

Theorem 1. *Let $n \geq 2$, and with every primitive, integral vector $\mathbf{a} \in R^n$ associate a measurable set $A(\mathbf{a})$ contained in the m-dimensional unit cube E_m. Then for almost all matrices ω there are either infinitely many or at most finitely many primitive integral vectors $\mathbf{a}$ which satisfy the condition*

$$(\mathbf{a}\cdot\bar{\omega}_1,\ \mathbf{a}\cdot\bar{\omega}_2,\ \ldots,\ \mathbf{a}\cdot\bar{\omega}_m) \in A(\mathbf{a}) \bmod 1, \tag{5}$$

depending on whether

$$\sum_{\mathbf{a}} |A(\mathbf{a})| < \infty, \tag{6}$$

or

$$\sum_{\mathbf{a}} |A(\mathbf{a})| = \infty \tag{7}$$

holds.

Theorem 2. *Under the conditions of the preceding theorem let $N(\omega, h)$ denote the number of primitive integral vectors $\mathbf{a} = (a_1, \cdots, a_n)$ with the property*

$$h(\mathbf{a}) = \max(|a_1|,\ |a_2|,\ \ldots,\ |a_n|) \leqslant h,$$

which satisfy the condition (5) for a fixed matrix ω. Furthermore, let $\Phi(h) = \Sigma_{h(\mathbf{a})\leq h} |A(\mathbf{a})|$ where the summation is extended over the primitive integral vectors $\mathbf{a}$. Then, for almost all ω, the equation

$$N(\omega,\ h) = \Phi(h) + O\left(\Phi^{\frac{1}{2}+\varepsilon}(h)\ln h\right) \tag{8}$$

is true.

The condition (5) may be cast into another form. Let Λ_0 be the lattice of all m-dimensional integral vectors and let $\Gamma = \Gamma(\omega)$ be the additive group generated by the column vectors $\omega^{(1)}, \cdots, \omega^{(n)}$ of the matrix (1). In other words, Γ is the set of all m-dimensional vectors $\gamma = a_1\omega^{(1)} + \cdots + a_n\omega^{(n)}$ with rational integer coefficients a_i. Then (5) is equivalent to the condition

$$a_1\omega^{(1)} + a_2\omega^{(2)} + \ldots + a_n\omega^{(n)} \in A(\mathbf{a}) \bmod \Lambda_0, \tag{9}$$

Thus the two preceding theorems are statements about the primitive points $\gamma \in \Gamma$ which belong to the factor space R^m/Λ_0. Instead of the actual lattice

Λ_0 we may consider any nonsingular lattice Λ, replacing the sets $A(\mathbf{a})$ from the m-dimensional unit cube by sets from a fundamental parallelepiped of the lattice (i.e. from the factor space R^m/Λ). It is evident that in the asymptotic formula analogous to (8) the principal term will then read $\Phi(h)d^{-1}(\Lambda)$, where $d(\Lambda)$ is the determinant of the lattice Λ.

We now turn to the proof of the Theorems 1 and 2.

Lemma 1. *The equations* (3) *and* (4) *are valid.*

Proof. We introduce in Ω_{1m} the "rectangle" $I = I_1 \times I_2 \times \cdots \times I_m$, where each I_j is an interval from E_1. Let χ_j be the characteristic function of the interval I_j, extended to the entire real line as a periodic function with period 1. Clearly we have

$$|T^{-1}I| = \int_{E_{mn}} \chi_1(\mathbf{a}\cdot\bar{\omega}_1)\dots\chi_m(\mathbf{a}\cdot\bar{\omega}_m)\,d\mu_{mn}, \tag{10}$$

where μ_{mn} denotes Lebesgue measure on E_{mn}. Let χ_j have the Fourier expansion

$$\chi_j(x) = a_{0j} + \sum_{\nu_j \neq 0} a_{\nu_j} e^{2\pi i \nu_j x}$$

Then it follows from (10) that

$$|T^{-1}I| = \int_{E_{mn}} a_{01}a_{02}\dots a_{0m}\,d\mu_{mn} + \dots, \tag{11}$$

where the terms, explicitly written, have the form

$$a_{\nu_1}a_{\nu_2}\dots a_{\nu_m}\int_{E_{mn}} \exp 2\pi i(\mathbf{a}\cdot\bar{\omega}_1\nu_1 + \dots + \mathbf{a}\cdot\bar{\omega}_m\nu_m)\,d\mu_{mn}, \tag{12}$$

with at least one of the indices $\nu_1, \nu_2, \cdots, \nu_m$ different from zero. Evidently,

$$\begin{aligned} &\mathbf{a}\cdot\bar{\omega}_1\nu_1 + \dots + \mathbf{a}\cdot\bar{\omega}_m\nu_m \\ &= \nu_1\sum_{j=1}^{n} a_j\omega_{1j} + \nu_2\sum_{j=1}^{n} a_j\omega_{2j} + \dots \\ &\dots + \nu_m\sum_{j=1}^{n} a_j\omega_{mj} = \sum_{i=1}^{m}\sum_{j=1}^{n}\nu_i a_j\omega_{ij}, \end{aligned}$$

and since $\mathbf{a} \neq (0)$, $\bar{\nu} = (\nu_1, \cdots, \nu_m) \neq (0)$, there exists a pair of indices

i, j with $\nu_i \neq 0$, $a_j \neq 0$. Using the identity

$$d\mu_{mn} = \prod_{\substack{1 \leq i \leq m \\ 1 \leq j \leq n}} d\omega_{ij}, \tag{13}$$

we thus obtain

$$\int_{E_{mn}} \exp 2\pi i (\mathrm{a}\cdot\bar{\omega}_1\nu_1 + \ldots + \mathrm{a}\cdot\bar{\omega}_m\nu_m)\, d\mu_{mn}$$

$$= \int_{E_{mn}} \exp 2\pi i \Big(\sum_{i,j} \nu_i a_j \omega_{ij}\Big)\, d\mu_{mn}$$

$$= \prod_{i,j} \int_{E_1} \exp(2\pi i \nu_i a_j \omega_{ij})\, d\omega_{ij} = 0.$$

Consequently, all the integrals (12) vanish. Finally, since

$$\alpha_{0j} = \int_{E_1} \chi_j(x)\, dx = |I_j|,$$

the equation (3) follows by means of (11).

The identity (3), which we have just proved for m-dimensional intervals, can can be generalized to Borel sets by the usual set-theoretical argument. Hence it can be shown to hold for all measurable sets A.

Now we come to the equation (4). Obviously, it is again sufficient to consider the case where the sets A_1 and A_2 are m-dimensional intervals. Thus, let

$$I^{(1)} = I_1^{(1)} \times \ldots \times I_m^{(1)}, \quad I^{(2)} = I_1^{(2)} \times \ldots \times I_m^{(2)}$$

be two such intervals in Ω_{1m} and let $\chi_j^{(1)}$ and $\chi_j^{(2)}$, respectively, denote the characteristic functions of the intervals $I_j^{(1)}$ and $I_j^{(2)}$, extended to the entire real line as periodic functions with period 1. Then we have

$$|T_1^{-1} I^{(1)} \cap T_2^{-1} I^{(2)}|$$

$$= \int_{E_{mn}} \prod_{j=1}^{m} \chi_j^{(1)}(\mathrm{a}\cdot\bar{\omega}_j) \prod_{j=1}^{m} \chi_j^{(2)}(\mathrm{a}\cdot\bar{\omega}_j)\, d\mu_m.$$

Expanding $\chi_j^{(1)}(x)$ and $\chi_j^{(2)}(x)$ as Fourier series, we obtain in analogy to the argument given above, the relation

$$|T_1^{-1} I^{(1)} \cap T_2^{-1} I^{(2)}| = \prod_{j=1}^{m} |I_j^{(1)}| \prod_{j=1}^{m} |I_j^{(2)}| + \dots, \tag{14}$$

where the terms of the sum have the form

$$\alpha_{\nu_1}^{(1)} \dots \alpha_{\nu_m}^{(1)} \alpha_{\lambda_1}^{(2)} \dots \alpha_{\lambda_m}^{(2)} \int_{E_{mn}} \exp 2\pi i$$

$$\times \left(\sum_{j=1}^{m} \mathbf{a}_1 \cdot \bar{\omega}_j \nu_j + \sum_{j=1}^{m} \mathbf{a}_2 \cdot \bar{\omega}_j \lambda_j\right) d\mu_{mn}, \tag{15}$$

with at least one of the indices $\nu_1, \dots, \nu_m, \lambda_1, \dots, \lambda_m$ different from zero. The sum under the integral sign is equal to

$$\sum_{j=1}^{m} \bar{\omega}_j \cdot (\mathbf{a}_1 \nu_j + \mathbf{a}_2 \lambda_j). \tag{16}$$

If (ν_j, λ_j) is a pair of indices with either $\nu_j \neq 0$ or $\lambda_j \neq 0$ then the presupposed linear independence of the vectors $\mathbf{a}_1, \mathbf{a}_2$ implies $\mathbf{a}_1\nu_j + \mathbf{a}_2\lambda_j \neq (0)$. Consequently, the scalar product (16) involves a term $b_{ij}\omega_{ij}$ in which b_{ij} is an integer different from zero. By (13) we may conclude that all the integrals (15) vanish. Consequently, (14) coincides with (4) in the case where A_1 and A_2 are m-dimensional intervals.

Lemma 2. *Let* $(\Omega, \mathbf{A}, \mu)$ *be a measure space, let a sequence of sets* $A_k \in \mathbf{A}$ *(measurable sets) be given such that*

$$\sum_{k=1}^{\infty} \mu A_k = \infty, \tag{17}$$

and let A *denote the set of points in* Ω *which belong to infinitely many sets* A_k. *Then the measure* μA *satisfies the inequality*

$$\mu A \geqslant \overline{\lim_{n \to \infty}} \frac{\left(\sum_{k=1}^{n} \mu A_k\right)^2}{\sum_{k,l=1}^{n} \mu (A_k \cap A_l)}. \tag{18}$$

Proof. Let $A^m = \bigcup_{k \geq m} A_k$. Then

$$A = \lim_{m\to\infty} \bigcup_{k\geqslant m} A_k = \lim_{m\to\infty} A^m.$$

Now let $A^n_m = \bigcup_{m \le k \le n} A_k$, and for each $\omega \in \Omega$ denote by $N^n_m(\omega)$ the number of those sets A_k, $m \le k \le n$, which contain ω. Thus we have, by the Cauchy-Schwarz-Bunjakovskiĭ Inequality

$$\Big(\int_\Omega N^n_m(\omega)\, d\mu\Big)^2 = \Big(\int_{A^n_m} N^n_m(\omega)\, d\mu\Big)^2$$

$$\leqslant \int_{A^n_m} (N^n_m(\omega))^2\, d\mu \cdot \int_{A^n_m} d\mu = \int_\Omega (N^n_m(\omega))^2\, d\mu \cdot \mu A^n_m$$

and consequently,

$$\mu A^n_m \geqslant \frac{\Big(\int_\Omega N^n_m(\omega)\, d\mu\Big)^2}{\int_\Omega (N^n_m(\omega))^2\, d\mu} \tag{19}$$

Now let χ_k be the characteristic function of the set A_k. Then

$$N^n_m(\omega) = \sum_{m\leqslant k\leqslant n} \chi_k(\omega);$$

hence we have

$$\int_\Omega N^n_m(\omega)\, d\mu = \sum_{m<k<n} \int_\Omega \chi_k(\omega)\, d\mu = \sum_{m\leqslant k<n} \mu A_k;$$

thus

$$\int_\Omega (N^n_m(\omega))^2\, d\mu = \sum_{m\leqslant k,l\leqslant n} \int_\Omega \chi_k(\omega)\chi_l(\omega)\, d\mu$$

$$= \sum_{m\leqslant k,l\leqslant n} \mu(A_k \cap A_l)$$

and consequently we obtain, by (19),

$$\mu A^n_m \geqslant \frac{\Big(\sum_{m\leqslant k\leqslant n} \mu A_k\Big)^2}{\sum_{m\leqslant k,l\leqslant n} \mu(A_k \cap A_l)}$$

Since $A^m \supseteq A^n_m$ for any $n \geq m$, it follows that

$$\mu A^m \geqslant \lim_{n\to\infty} A^n_m \geqslant \overline{\lim_{n\to\infty}} \frac{\left(\sum_{m\leqslant k\leqslant n} \mu A_k\right)^2}{\sum_{m\leqslant k,l\leqslant n} \mu (A_k \cap A_l)}.$$

Finally, (17) implies the asymptotic equations

$$\sum_{m\leqslant k<n} \mu A_k \sim \sum_{1\leqslant k\leqslant n} \mu A_k \quad (n \to \infty),$$

$$\sum_{m\leqslant k,l\leqslant n} \mu (A_k \cap A_l) \sim \sum_{1\leqslant k,l\leqslant n} \mu (A_k \cap A_l) \quad (n \to \infty).$$

Hence μA^m is not smaller than the right-hand side of (18), and since $\mu A = \lim_{m\to\infty} \mu A^m$, the proof of the inequality (18) is completed.

Lemma 3. *Let ξ_k $(k = 1, 2, \cdots)$ be mutually orthogonal random variables with expectation zero and with finite variance $B_n = \Sigma_{k\leq n} E|\xi_k|^2$. Then the asymptotic equation*

$$\sum_{k=1}^{n} \xi_k = O\left(B_n^{\frac{1}{2}+\varepsilon} \ln n\right) \tag{20}$$

holds with probability one.

Proof. This lemma is an immediate consequence of well-known results on orthogonal random variables. It is known, for example, that for any sequence (b_n) of numbers which increases monotonically to infinity and satisfies the condition

$$\sum_{k=1}^{\infty} \left(\frac{\ln n}{b_n}\right)^2 E|\xi_k|^2 < \infty, \tag{21}$$

the asymptotic equation $\Sigma_{k=1}^n \xi_k = o(b_n)$ holds with probability 1 (see [39], Chapter X, §33.1). If we choose

$$b_n = B_n^{\frac{1}{2}+\varepsilon} \ln n,$$

then it is easy to show that (21) and, consequently, (20) are true. Indeed, the convergence of the series (21) is, under the present conditions, an immediate consequence of the following remark on series of nonnegative numbers: Let $a_k \geq 0$ $(k = 1, 2, \cdots)$ be real numbers, $S_n = \Sigma_{k=1}^n a_k$; then the series

$$\sum_{n=1}^{\infty} a_n S_n^{-1-\varepsilon} \tag{22}$$

converges for any given $\epsilon > 0$.

The proof is simple. On one hand we have

$$a_{n+1} \int_n^{n+1} (S_n + (x-n)\, a_{n+1})^{-1-\varepsilon}\, dx \geqslant a_{n+1} S_{n+1}^{-1-\varepsilon}.$$

On the other hand, if $a_{n+1} \neq 0$, then

$$a_{n+1} \int_n^{n+1} (S_n + (x-n)\, a_{n+1})^{-1-\varepsilon}\, dx = \frac{1}{\varepsilon} (S_n^{-\varepsilon} - S_{n+1}^{-\varepsilon}).$$

Consequently,

$$a_{n+1} S_{n+1}^{-1-\varepsilon} \leqslant \frac{1}{\varepsilon} (S_n^{-\varepsilon} - S_{n+1}^{-\varepsilon}),$$

which implies the convergence of the series (22).

Proof of Theorem 1. In the case where (6) holds, the reasoning is obvious. In fact, let Ω'_{mn} be the set of those $\omega \in \Omega_{mn}$ for which (5) is satisfied infinitely often. Then we have, for any h,

$$\Omega'_{mn} \subseteq \bigcup_{h(\mathbf{a}) \geqslant h} T_{\mathbf{a}}^{-1} A(\mathbf{a})$$

Consequently, by (3) and (6),

$$|\Omega'_{mn}| \leqslant \lim_{h \to \infty} \sum_{h(\mathbf{a}) \geqslant h} |T_{\mathbf{a}}^{-1} A(\mathbf{a})| = \lim_{h \to \infty} \sum_{h(\mathbf{a}) \geqslant h} |A(\mathbf{a})| = 0.$$

Now we admit any pair $\mathbf{a}_1, \mathbf{a}_2$ of primitive integral vectors with $\mathbf{a}_1 \neq \pm \mathbf{a}_2$. Then they are linearly independent. Otherwise, let $\mathbf{a}_1 \nu_1 + \mathbf{a}_2 \nu_2 = 0$, where ν_1, ν_2 are integers. We may assume that $(\nu_1, \nu_2) = 1$. Then $\nu_1 | \mathbf{a}_2 \nu_2$ and thus $\nu_1 | \mathbf{a}_2$. Analogously, $\nu_2 | \mathbf{a}_1$. But since the vectors $\mathbf{a}_1, \mathbf{a}_2$ are primitive, this implies $\nu_1 = \pm 1$, $\nu_2 \neq \pm 1$, and hence $\mathbf{a}_1 = \pm \mathbf{a}_2$, contrary to our assumption.

In order to prove the theorem in the case where (7) holds, we apply Lemma 2 to a sequence of sets $T_{\mathbf{a}}^{-1} A(\mathbf{a})$ (with primitive vectors $\mathbf{a}$), numbered in the order of increasing values $h(\mathbf{a})$.

By applying the equations (3) and (4) we obtain

$$\sum_{h(\mathbf{a}_1),\, h(\mathbf{a}_2)\leqslant h} |T_{\mathbf{a}_1}^{-1}A(\mathbf{a}_1)\cap T_{\mathbf{a}_2}^{-1}A(\mathbf{a}_2)| = \sum_{\mathbf{a}_1\neq\mathbf{a}_2} + \sum_{\mathbf{a}_1=\pm\mathbf{a}_2}$$
$$\leqslant \sum_{\mathbf{a}_1\neq\mathbf{a}_2} |A(\mathbf{a}_1)|\,|A(\mathbf{a}_2)| + 2\sum_{\mathbf{a}} |A(\mathbf{a})|$$
$$< \Big(\sum_{\mathbf{a}} |A(\mathbf{a})|\Big)^2 + 2\sum_{\mathbf{a}} |A(\mathbf{a})|$$
$$= (1+o(1))\Big(\sum_{\mathbf{a}} |A(\mathbf{a})|\Big)^2,$$

where the symbol $o(1)$ stands for a quantity which converges to zero as h tends to infinity.

By Lemma 2 this implies $|\Omega'_{mn}| = 1$, thus completing the proof of Theorem 1.

Proof of Theorem 2. Let $\chi_{\mathbf{a}}(\omega)$ be the characteristic function of the set $T_{\mathbf{a}}^{-1}A(\mathbf{a})$. Then we have, by (3),

$$E\chi_{\mathbf{a}} = \int_{E_{mn}} \chi_{\mathbf{a}}(\omega)\, d\mu_{mn} = |T_{\mathbf{a}}^{-1}A(\mathbf{a})| = |A(\mathbf{a})|.$$

We define the random variables $\xi_{\mathbf{a}}(\omega) = \chi_{\mathbf{a}}(\omega) - |A(\mathbf{a})|$. By (4), this implies for linearly independent vectors $\mathbf{a}_1$, $\mathbf{a}_2$ the relation

$$E\xi_{\mathbf{a}_1}\xi_{\mathbf{a}_2} = \int_{E_{mn}} (\chi_{\mathbf{a}_1}(\omega) - |A(\mathbf{a}_1)|)(\chi_{\mathbf{a}_2}(\omega) - |A(\mathbf{a}_2)|)\, d\mu_{mn} =$$
$$= \int_{E_{mn}} \chi_{\mathbf{a}_1}(\omega)\chi_{\mathbf{a}_2}(\omega)\, d\mu_{mn} - |A(\mathbf{a}_1)|\,|A(\mathbf{a}_2)|$$
$$= |T_{\mathbf{a}_1}^{-1}A(\mathbf{a}_1)\cap T_{\mathbf{a}_2}^{-1}A(\mathbf{a}_2)| - |A(\mathbf{a}_1)|\,|A(\mathbf{a}_2)| = 0.$$

Consequently, the random variables $\xi_{\mathbf{a}_1}$, $\xi_{\mathbf{a}_2}$ are orthogonal whenever the vectors $\mathbf{a}_1$, $\mathbf{a}_2$ are linearly independent.

We subdivide the set of all n-dimensional primitive vectors $\mathbf{a}$ into classes by placing into the classes M_k^+ and M_k^-, respectively, all vectors $\mathbf{a}$ with $a_1 = a_2 = \cdots = a_{k-1} = 0$ for which $a_k > 0$ or $a_k < 0$. Obviously, any two sets so defined are disjoint if either the indices k or the $+$ and $-$ signs involved are different. In each class, any pair $\mathbf{a}_1$, $\mathbf{a}_2$ of vectors is linearly independent. Consequently, considering the random variables $\xi_{\mathbf{a}}$ for all $\mathbf{a} \in M_k^+$ or for all $\mathbf{a} \in M_k^-$, we can apply Lemma 3. For example, for the class M_k^+ this yields the relation

$$\sum_{\substack{a \in M_k^+ \\ h(a) \leqslant h}} (\chi_a(\omega) - |A(a)|) = O(B_n^{\frac{1}{2}+\varepsilon} \ln n),$$

where

$$B_n = \sum_{\substack{a \in M_k^+ \\ h(a) \leqslant h}} E|\xi_a|^2 < \sum_{h(a) \leqslant h} |A(a)|,$$

since $E|\xi_a|^2 = |A(a)| - |A(a)|^2$. Clearly, there are $2n$ such classes, and, since they are mutually disjoint, we have

$$\sum_{h(a) \leqslant h} (\chi_a(\omega) - |A(a)|) = O\left(\left(\sum_{h(a) \leqslant h} |A(a)|\right)^{\frac{1}{2}+\varepsilon} \ln h\right),$$

which is precisely the assertion of Theorem 2.

A number of interesting corollaries can be derived from Theorems 1 and 2. For example, let M be an arbitrary set of primitive integral vectors $\mathbf{a} \in R^n$, and let $\phi(\mathbf{a})$ be a nonnegative function, defined on M, with $\sum_{\mathbf{a} \in M} \phi(\mathbf{a}) = \infty$. Furthermore let $\{x\}$ denote the fractional part of x. Then the inequality

$$\{a_1 \omega_1 + a_2 \omega_2 + \ldots + a_n \omega_n\} < \varphi(\mathbf{a}), \quad \mathbf{a} \in M$$

has, for almost all $\omega = (\omega_1, \omega_2, \cdots, \omega_n)$, infinitely many solutions $\mathbf{a}$. In particular, this is true if we choose as M the set of all n-tuples of the form $(a_1^s, a_2^s, \cdots$ $\cdots, a_n^s)$, where s is a natural number, $(a_1, a_2, \cdots, a_n) = 1$, $\phi(\mathbf{a}) = h^{-n}$, $h = $ max $(|a_1|, |a_2|, \cdots, |a_n|)$. Consequently, for almost all ω there exist infinitely many integral vectors $(a_1, a_2, \cdots, a_n)$ which satisfy the inequality

$$\{a_1^s \omega_1 + a_2^s \omega_2 + \ldots + a_n^s \omega_n\} < h^{-n}$$

An analogous assertion is true for the inequality

$$\{g_1^l \omega_1 + g_2^l \omega_2 + \ldots + g_n^l \omega_n\} < l^{-1},$$

where the g_i and l are natural numbers with $(g_1, g_2, \cdots, g_n) = 1$. In reality, Theorem 1 gives results of this type for *systems* of inequalities, whereas Theorem 2 describes the asymptotic number of solutions of such inequalities. Clearly it is possible in a given case to derive statements on "nonlinear" diophantine approximations, since in Theorems 1 and 2, even though they are concerned with linear diophantine approximations, the choice of the sets $A(\mathbf{a})$ is subject only to the condition (7), and thus we can deal with nonlinear problems by taking the empty set as $A(\mathbf{a})$ for suitable values of $\mathbf{a}$.

§2. SIMULTANEOUS APPROXIMATIONS OF NUMBERS WITH A QUADRATIC RELATION

Let $n \geq 2$ be an arbitrary (but fixed) integer and let $\|x\|$ denote the distance between x and the nearest integer.

Theorem. *For each n-tuple* $\bar{\omega} = (\omega_1, \omega_2, \cdots, \omega_n)$ *of real numbers let* $w(\bar{\omega})$ *denote the supremum of the set of those numbers* $w > 0$ *for which there exist infinitely many polynomials*

$$\Phi(x_1, x_2, \ldots, x_n) = \sum_{1 \leqslant i,\, j \leqslant n} a_{ij} x_i x_j + \sum_{1 \leqslant i \leqslant n} a_i x_i \quad (a_{ij} = a_{ji}),$$

with integer coefficients, such that the inequality

$$\|\Phi(\omega_1, \omega_2, \ldots, \omega_n)\| < h^{-w}, \quad h = \max_{(i,j)} (|a_{ij}|, |a_i|)$$

is satisfied. Analogously, let $\lambda(\bar{\omega})$ *denote the supremum of the set of those* $\lambda > 0$ *for which the system of inequalities*

$$\max_{(i,j)} (\|\omega_i q\|, \|\omega_i \omega_j q\|) < q^{-\lambda} \quad (i, j = 1, 2, \ldots, n)$$

has infinitely many integers $q > 0$ *as solutions. Then the equations*

$$w(\bar{\omega}) = \frac{n(n+3)}{2} \quad (n = 2, 3, \ldots), \tag{1}$$

$$\lambda(\bar{\omega}) = \frac{2}{n(n+3)} \quad (n = 2, 3, \ldots) \tag{2}$$

hold for almost all $\bar{\omega}$.

Proof. Evidently, the equations (1) and (2) are equivalent by Hinčin's principle of transfer. In order to prove the theorem it suffices to show that $\lambda(\bar{\omega}) \leq 2/n(n+3)$ for almost all $\bar{\omega}$, since the converse inequality follows immediately from Minkowski's lemma on linear forms.

Thus we have to prove that, for almost all $\bar{\omega}$, there exist at most finitely many integers $q > 0$ which solve the system

$$\max_{(i,j)} (\|\omega_i q\|, \|\omega_i \omega_j q\|) < q^{-\frac{1}{m} - \delta} \quad (i, j = 1, 2, \ldots, n) \tag{3}$$

of inequalities, where $m = n(n+3)/2$ and δ is an arbitrary positive number.

We may restrict our consideration to vectors $\overline{\omega} = (\omega_1, \cdots, \omega_n)$ which satisfy the conditions

$$0 \leq \omega_i < 1 \quad (i = 1, 2, \cdots, n). \tag{4}$$

Indeed, let $\omega_i = \omega'_i + l_i$ $(i = 1, \cdots, n)$, with integers l_i. Then, using the trivial properties

$$\|x\| = \|-x\|, \quad \|x+y\| \leq \|x\| + \|y\|,$$

we have the relations

$$\|\omega_i q\| = \|\omega'_i q\| \quad (i = 1, 2, \ldots, n),$$

$$\|\omega_i\omega_j q\| = \|\omega'_i \omega'_j q + (l_i \omega'_j + l_j \omega'_i) q\|$$

$$\leqslant \|\omega'_i \omega'_j q\| + |l_i| \, \|\omega'_j q\| + |l_j| \, \|\omega'_i q\| \quad (i, j = 1, 2, \ldots, n).$$

Thus there exists a number $L = L(\overline{\omega})$, independent of q, such that

$$\max_{(i,j)} (\|\omega_i q\|, \|\omega_i\omega_j q\|) \leqslant L \max_{(i,j)} (\|\omega'_i q\|, \|\omega'_i \omega'_j q\|).$$

It is obvious that, if necessary, the integers l_i can be chosen in such a way that the numbers ω'_i satisfy the conditions (4).

Turning to the investigation of the system (3), we introduce the notation

$$\|\omega_i q\| = |\omega_i q - a_i|, \quad \|\omega_i\omega_j q\| = |\omega_i\omega_j q - a_{ij}|$$

$$(i, j = 1, 2, \ldots, n). \tag{5}$$

Then (3) implies the inequalities

$$\left|\omega_i - \frac{a_i}{q}\right| < q^{-1-\frac{1}{m}-\delta}, \tag{6}$$

$$\left|\omega_i\omega_j - \frac{a_i a_j}{q}\right| < (|a_i| + |a_j|) q^{-2-\frac{1}{m}-\delta} + q^{-2-\frac{2}{m}-2\delta} \tag{7}$$

for all i, j. By (4) and (6) we have $a_i = \omega_i q + \Theta_i q^{-1/m-\delta}$, $|\Theta_i| < 1$, and thus $-q^{-1/m-\delta} < a_i < q + q^{-1/m-\delta}$. But since the a_i are integers, it follows that

$$0 \leq a_i \leq q \quad (i = 1, 2, \cdots, n). \tag{8}$$

Consequently, from (3), (5) and (7) we have the inequality $|a_i a_j / q - a_{ij}| < 4q^{-1/m-\delta}$ or, in another notation,

$$\left\|\frac{a_i a_j}{q}\right\| < 4q^{-\frac{1}{m}-\delta} \qquad (i,\ j = 1,\ 2,\ \ldots,\ n). \tag{9}$$

Our further efforts will be directed towards proving the estimate

$$N(q) \ll q^{n+\frac{n}{m}-1}\, 2^{\alpha(q)}, \tag{10}$$

for the number of solutions of the system (9) of inequalities under restriction (8). Here $\alpha(q)$ is the greatest exponent for which $2^{\alpha} | q$. Once this inequality has been established, it is obvious how we have to proceed further: For a fixed n-tuple $(a_1, a_2, \cdots, a_n)$ the set of points $\bar{\omega}$ satisfying the condition (6) has measure $(2q^{-1-1/m-\delta})^n$, and the number of possible n-tuples for a fixed q is bounded by the inequality (10). Hence the measure of the set of points ω satisfying the system (3) for a given q is

$$\ll q^{-n-\frac{n}{m}-\delta n}\, q^{n+\frac{n}{m}-1} \cdot 2^{\alpha(q)} = q^{-1-\delta_1} \cdot 2^{\alpha(q)}.$$

Furthermore we have

$$\sum_{q=1}^{\infty} q^{-1-\delta_1}\, 2^{\alpha(q)} \leqslant \sum_{\alpha=0}^{\infty} \sum_{q_1=1}^{\infty} (2^{\alpha} q_1)^{-1-\delta_1} \cdot 2^{\alpha}$$

$$= \sum_{\alpha=0}^{\infty} 2^{-\alpha\delta_1} \sum_{q_1=1}^{\infty} q_1^{-1-\delta_1} = (1-2^{-\delta_1})^{-1} \zeta(1+\delta_1) < \infty,$$

and thus the assertion of the theorem follows.

We will prove the inequality (10) by means of known properties of the Gaussian sums

$$\Gamma_q(S) = \sum_{(a_1, a_2, \ldots, a_n) \bmod q} \exp 2\pi i \frac{S(a_1,\ a_2,\ \ldots,\ a_n)}{q},$$

associated with the integral quadratic forms $S(a_1, \cdots a_n)$.

Let the numbers c_{ij} be arbitrary integers satisfying

$$c_{ij} = c_{ji},\quad |c_{ij}| \le c \quad (i,\ j = 1,\ 2,\ \cdots,\ n). \tag{11}$$

Applying the elementary inequality $|e^{2\pi i x} - 1| \le 2\pi \|x\|$ and using the notation

$$S(a_1,\ a_2,\ \ldots,\ a_n) = \sum_{1 \leqslant i,\ j \leqslant n} c_{ij} a_i a_j, \tag{12}$$

we obtain by (9) the relation

$$\left| \exp 2\pi i \frac{S(a_1, a_2, \ldots, a_n)}{q} - 1 \right| \leqslant 2\pi \left\| \frac{S(a_1, a_2, \ldots, a_n)}{q} \right\|$$

$$\leqslant 2\pi \sum_{1 \leqslant i,\, j \leqslant n} |c_{ij}| \left\| \frac{a_i a_j}{q} \right\| < 8\pi n^2 c q^{-\frac{1}{m} - \delta} < \frac{1}{2},$$

where

$$c = [(16\pi n^2)^{-1} q^{\frac{1}{m}}]. \tag{13}$$

Thus, if the vector a $= (a_1, a_2, \cdots, a_n)$ **satisfies the system (9) of inequalities by summing over all integral quadratic forms (12) with the properties described by (11) and (13), we have the relation**

$$\left| \sum_S \exp 2\pi i \frac{S(a_1, \ldots, a_n)}{q} \right| > \frac{1}{2} \sum_S 1 \gg q^{\frac{m_1}{m}},$$

where $m_1 = n(n+1)/2$. Summing the last inequality over all $\mathbf{a} = (a_1, a_2, \cdots, a_n)$ which satisfy (8), we obtain

$$\sum_{\mathbf{a}} \left| \sum_S \exp 2\pi i \frac{S(\mathbf{a})}{q} \right|^2 \gg N(q) q^{\frac{2m_1}{m}}.$$

On the other hand, denoting the summation on the left-hand side by Σ, we have

$$\sum_{\mathbf{a}} \sum_{S_1} \sum_{S_2} \exp 2\pi i \frac{S_1(\mathbf{a}) - S_2(\mathbf{a})}{q}$$

$$\leqslant N\{S_1 - S_2 = S\} \left| \sum_{\mathbf{a}} \exp 2\pi i \frac{S(\mathbf{a}_1)}{q} \right|,$$

where $N\{S_1 - S_2 = S\}$ denotes the number of representations of the quadratic form S as the difference of two forms S_1, S_2 satisfying (11) and (13). It is evident that $N\{S_1 - S_2 = S\} \ll q^{m_1/m}$ and that the height $h(S)$ of the form S does not exceed $2c$. Thus we have

$$\Sigma \ll q^{\frac{m_1}{m}} \sum_S \left| \sum_{\mathbf{a}} \exp 2\pi i \frac{S(\mathbf{a})}{q} \right|,$$

which implies

$$N(q) \ll q^{-\frac{m_1}{m}} \sum_S \left| \sum_{\mathbf{a}} \exp 2\pi i \frac{S(\mathbf{a})}{q} \right|.$$

Treating separately the case where all coefficients of the form S are equal to zero and where at least one of the components of a is equal to q, we obtain

$$N(q) \ll q^{-\frac{m_1}{m}+n} + q^{-\frac{m_1}{m}} \sum_{S \neq (0)} |\Gamma_q(S)|, \tag{14}$$

where $\Gamma_q(S)$ denotes the Gaussian sum of the quadratic form S and the summation extends over all nonsingular quadratic forms whose heights $h(S)$ satisfy the condition

$$0 \neq h(S) = \max_{(i,j)} |c_{ij}| \leqslant 2c \quad (i,\ j = 1, 2, \ldots, n). \tag{15}$$

Thus we have to find an estimate for the summation

$$\sum\nolimits_q = \sum_{S \neq (0)} |\Gamma_q(S)|. \tag{16}$$

For a fixed form S' we denote by $d(S')$ the greatest common divisor of all its coefficients and of the modulus q involved, i.e. $d = (\ldots, c_{ij}, \ldots q)$. Then

$$\Gamma_q(S) = \sum_{\mathbf{a} \bmod q} \exp 2\pi i \frac{S(\mathbf{a})}{q}$$

$$= \sum_{l_1=0}^{d-1} \cdots \sum_{l_n=0}^{d-1} \sum_{\mathbf{a}' \bmod q_1} \exp 2\pi i \frac{S_1(l_1 d + a'_1, \ldots, l_n d + a'_n)}{q_1}$$

$$= d^n \sum_{\mathbf{a} \bmod q_1} \exp 2\pi i \frac{S_1(\mathbf{a})}{q_1} = d^n \Gamma_{q_1}(S_1),$$

where $S_1 = d^{-1}S$ is an integral form whose coefficients, together with the number $q_1 = d^{-1}q$, form a relatively prime system. By grouping together all forms S with the same number $d = d(S)$ we obtain for the summation (16) the inequality

$$\sum\nolimits_q \leqslant \sum_{d|q} d^n \sum_{0 \neq h(S_1) \leqslant 2cd^{-1}} |\Gamma_{q_1}(S_1)|, \tag{17}$$

where the inner summation is taken over all nonsingular forms S_1 whose coefficients are prime to the modulus q_1 and whose heights do not exceed $2cd^{-1}$ with c defined as in (13).

In order to obtain an estimate for the sum $\Gamma_{q_1}(S_1)$ occurring in (17) we apply the well-known procedure of reducing a Gaussian sum to canonical form.

Let $q_1 = s_1 s_2 \cdots s_r$, where the numbers s_i are mutually relatively prime, let $M_i = q_1/s_i$ $(i = 1, 2, \cdots, r)$; furthermore let integers M'_i be defined by the congruences

$$M_i M_i' \equiv 1 \pmod{s_i} \quad (i = 1, 2, \ldots, r) \tag{18}$$

and let $E_i = M_i M'_i$ $(i = 1, 2, \cdots, r)$. Then, obviously,

$$E_i E_j \equiv \begin{cases} E_i, & i = j \\ 0, & i \neq j \end{cases} \pmod{q_1}, \tag{19}$$

since there exists a one-to-one correspondence between the complete system of residue classes $X \pmod{q_1}$ and the direct product of the complete systems of residues $x_1 \pmod{m_1}, \cdots, x_r \pmod{m_r}$; such a mapping

$$X \pmod{q_1} \longleftrightarrow x_1 \pmod{m_1}, \ \ldots, \ x_r \pmod{m_r}$$

is accomplished by the congruence

$$X \equiv E_1 x_1 + \ldots + E_r x_r \pmod{q_1}. \tag{20}$$

In order to describe this mapping, we introduce the notation

$$a_i \equiv E_1 a_{1i} + \ldots + E_r a_{ri} \pmod{q_1} \quad (i = 1, 2, \ldots, n).$$

Thus we have, by (19),

$$a_i a_j \equiv \sum_{k,l=1}^{r} E_k E_l a_{ki} a_{lj} \equiv \sum_{k=1}^{r} E_k a_{kl} a_{kj} \pmod{q_1}.$$

Consequently, switching from these congruences to the corresponding equations, performing obvious algebraic operations and finally equating fractional parts, we obtain

$$\left\{ \sum_{i,j} \frac{c_{ij} a_i a_j}{q_1} \right\} = \left\{ \sum_{i,j} \frac{c_{ij} E_1 a_{1i} a_{1j}}{q_1} + \ldots + \sum_{i,j} \frac{c_{ij} E_r a_{ri} a'_{rj}}{q_1} \right\}$$

$$= \left\{ \sum_{i,j} \frac{c_{ij} M'_1 a_{1i} a_{1j}}{m_1} + \ldots + \sum_{i,j} \frac{c_{ij} M'_r a_{ri} a_{rj}}{m_r} \right\}.$$

On the other hand, since a_i and a_j run through complete systems of residues modulo q_1, it follows that the numbers a_{ki} and a_{kj} run through complete systems of residues modulo m_k $(k = 1, 2, \cdots, r)$.

We apply this remark in the case where $s_1 = 2^\alpha$ with $2^\alpha \| q_1$ (i.e. 2 is an exact divisor of q_1) and $s_2 = Q = q_1 2^{-\alpha}$. This yields the relation

$$\Gamma_{q_1}(S_1) = \sum_{\mathbf{a}_1 \bmod s_1} \exp 2\pi i \frac{M_1 S_1(\mathbf{a}_1)}{s_1} \sum_{\mathbf{a}_2 \bmod s_2} \exp 2\pi i \frac{M_2' S_1(a_2)}{s_2}$$

$$= \Gamma_{s_1}(M_1' S_1) \Gamma_{s_2}(M_2' S_1),$$

and therefore

$$|\Gamma_{q_1}(S_1)| = |\Gamma_{2^\alpha}(M_1' S_1)| \, |\Gamma_Q(M_2' S_1)| \leqslant 2^{\alpha n} |\Gamma_Q(S_2)|, \tag{21}$$

where $Q = q_1 2^{-\alpha}$ is an odd integer and $S_2 = M'_2 S_1$ is an integral quadratic form whose coefficients are jointly relatively prime to Q.

Let p be any prime divisor of Q with $p^\rho \| Q$. Then it is easy to find a linear transformation $\mathbf{X} = A^{(p)}\mathbf{x}$, with integer coefficients, satisfying $(\det A^{(p)}, p) = 1$, for which a congruence of the form

$$S_2(\mathbf{x}) \equiv f_1^{(p)} X_1^2 + S_3^{(p)}(X_2, \ldots, X_n) \pmod{p^\rho}, \tag{22}$$

holds identically in $\mathbf{x}$ and $\mathbf{X}$, with $(f_1^{(p)}, p) = 1$. In order to change from this congruence to a congruence modulo Q we introduce a transformation of natural numbers induced by the decomposition (20). Let Q have the canonical representation $Q = p_1^{\rho_1} \cdots p_r^{\rho_r}$. Then we define for the system of moduli $p_1^{\rho_1}, \cdots, p_r^{\rho_r}$ numbers $E_1, \cdots, E_r$ by (18) and (19), letting $s_i = p_i^{\rho_i}$ $(i = 1, 2, \cdots, r)$. If $u_1, \cdots, u_r$ are residues mod $p_1^{\rho_1}, \cdots, p_r^{\rho_r}$, respectively, then we define the number u by the equation

$$u = E_1 u_1 + \cdots + E_r u_r. \tag{23}$$

Let $A^{(p)} = (\alpha_{ij}^{(p)})$ $(i, j = 1, 2, \cdots, n)$. Applying (23), we let

$$\alpha_{ij} = \sum_{p|Q} E^{(p)} \alpha_{ij}^{(p)}, \quad f_1 = \sum_{p|Q} E^{(p)} f_1^{(p)}, \tag{24}$$

where $E^{(p_i)} = E_i$ $(i = 1, 2, \cdots, r)$. Now we perform over the residue-class ring mod Q the invertible transformation $\mathbf{X} = A\mathbf{x}$, where $A = (\alpha_{ij})$, the α_{ij} being defined by (24). From the forms $S_3^{(p)}$ we construct a form S_3 whose coefficients are obtained from the coefficients of these forms by equations analogous to (24), again applying (23). Then we obtain the congruence

$$S_2(\mathbf{x}) \equiv f_1 X_1^2 + S_3(X_2, \ldots, X_n) \pmod{Q}, \tag{25}$$

where clearly $(f_1, Q) = 1$, and hence in turn we get the equation

$$\Gamma_Q(S_2) = \sum_{X_1 \bmod Q} \exp 2\pi i \frac{f_1 X_1^2}{Q}$$

$$\times \sum_{(X_2, \ldots, X_n) \bmod Q} \exp 2\pi i \frac{S_3(X_3, \ldots, X_n)}{Q}.$$

Furthermore, if Q_2 is the greatest common divisor of the coefficients of the form S_3 and of the number Q, then we have

$$|\Gamma_Q(S_2)| = Q^{\frac{1}{2}} Q_2^{n-1}$$

$$\times \left| \sum_{(x_2, \ldots, x_n) \bmod q_2} \exp 2\pi i \frac{S_3'(x_2, \ldots, x_n)}{q_2} \right|, \tag{26}$$

with $q_2 = Q/Q_2$ and with the form S'_3 being primitive relative to q_2. Here we have made use of the well-known identity (cf. [70], Chapter V, problem 11)

$$\left| \sum_{x=0}^{q-1} \exp 2\pi i \frac{ax^2}{q} \right| = q^{\frac{1}{2}}, (a, q) = 1, \quad q \equiv 1 \pmod 2. \tag{27}$$

The preceding argum... to the form S'_3, thus leading to a decomposition of the type (25) modulo q_2 which in turn yields an equation of the type (26), etc. Applying (27) at each step, we continue this procedure until we have obtained a quadratic form in one variable or until we have reached a module $q_{k+1} = 1$. Hence we have

$$|\Gamma_Q(S_2)| = Q^{\frac{1}{2}} Q_2^{n-1} q_2^{\frac{1}{2}} Q_3^{n-2} \ldots q_{k-1}^{\frac{1}{2}} Q_k^{n-k+1} q_k^{\frac{1}{2}} Q_{k+1}^{n-k},$$

where $Q = q_2 Q_2$, $q_2 = q_3 Q_3, \cdots, q_{k-1} = q_k Q_k$, $Q_{k+1} = q_k$, $k \leq n$. Combining these equations, we find $Q = Q_2 Q_3 \cdots Q_{k+1}$; consequently,

$$Q_2^{n-1} Q_3^{n-2} \ldots Q_{k+1}^{n-k} = \frac{Q^n}{Q_2 Q_3^2 \ldots Q_{k+1}^k} = \frac{Q^{n-1}}{q_2 q_3 \cdots q_k}.$$

and thus

$$|\Gamma_Q(S_2)| = \frac{Q^{n-\frac{1}{2}}}{(q_2q_3\cdots q_k)^{\frac{1}{2}}} \ll Q^{n-\frac{1}{2}} q_2^{-\frac{1}{2}} = Q^{n-1} Q_2^{\frac{1}{2}}. \tag{28}$$

It follows from (25) that the congruence

$$M_2' S_1(\mathbf{x}) \equiv f_1 X_1^2 \pmod{Q_2},$$

holds identically in $\mathbf{x}$ if, say, $X_1 = \alpha_1 x_1 + \alpha_2 x_2 + \cdots + \alpha_n x_n$; furthermore, $(M_2', Q_2) = 1$. Consequently, the coefficients c_{ij} of the form S_1 satisfy the congruences

$$M_2' c_{ij}' \equiv f_1 \alpha_i \alpha_j \pmod{Q_2},$$

$$c_{ij}' \equiv g\,\alpha_i \alpha_j \pmod{Q_2}, \tag{29}$$

where g is an integer with $(g, Q_2) = 1$. We consider Q_2 as a fixed divisor of Q and derive a lower bound for the number of quadratic forms S_1 satisfying (29) under the condition

$$|c_{ij}'| \leqslant 2cd^{-1} \quad (i, j = 1, 2, \ldots, n).$$

From (29) we have

$$\alpha_i^2 \equiv g' c_{ii}' \pmod{Q_2}, \quad gg' \equiv 1 \pmod{Q_2}. \tag{30}$$

We know that the number of solutions of the congruence $x^2 \equiv a \pmod{m}$ is a quantity $\ll m^{\epsilon}(a, m)^{1/2}$ (see Part II, Chapter 2, §5). Consequently, allowing the coefficients $c_{11}', c_{22}', \cdots, c_{nn}'$ to vary independently, we find that the number of possible choices of the vector $\overline{\alpha} = (\alpha_1, \alpha_2, \cdots, \alpha_n) \bmod Q_2$ is

$$\ll \prod_{i=1}^{n} Q_2^{\varepsilon}(c_{ii}', Q_2)^{\frac{1}{2}} \tag{31}$$

If $Q_2 > 4cd^{-1}$ then it follows from (29) that the coefficients c'_{ij} are uniquely determined by $\overline{\alpha}$ as the residue classes mod Q_2 represented by the numbers $g\alpha_i\alpha_j$ $(1 \leq i < j \leq n)$. If $Q_2 \leq 4cd^{-1}$ then the coefficients c'_{ij} for a given $\overline{\alpha}$ form an arithmetical progression with the difference Q_2 in which the number of terms is $\ll cd^{-1}Q_2^{-1}$. Therefore, expressing c by (13), we find that for fixed α_i and α_j the number of possible values of c'_{ij} is in both cases $\ll 1 + q^{1/m}d^{-1}Q_2^{-1}$, and hence the total number of possible values of c'_{ij} $(1 \leq i < j \leq n)$ for a given

$\bar{\alpha}$ is $\ll (1 + q^{1/m} d^{-1} Q_2^{-1})^{m_2}$, $m_2 = n(n-1)/2$.

Applying the estimation (31) on the number of possible choices of the vector $\bar{\alpha}$ for given values of the numbers c'_{ij} $(i = 1, 2, \cdots, n)$ we then find that the total number of possible values of the coefficients c'_{ij} is

$$\ll \sum_{|c'_{11}| \ll q^{\frac{1}{m}} d^{-1}} \cdots \sum_{|c'_{nn}| \ll q^{\frac{1}{m}} d^{-1}} \prod_{i=1}^{n} Q_2^{\varepsilon} (c'_{ii}, Q_2)^{\frac{1}{2}}$$

$$\times (1 + q^{\frac{1}{m}} d^{-1} Q_2^{-1})^{m_2}$$

$$< Q_2^{\varepsilon n} (1 + q^{\frac{m_2}{m}} d^{-m_2} Q_2^{-m_2}) \Big(\sum_{\sigma \ll q^{\frac{1}{m}} d^{-1}} (\sigma, Q_2)^{\frac{1}{2}} \Big)^n$$

and since we have

$$\sum_{\sigma \leqslant T} (\sigma, Q_2)^{\frac{1}{2}} < \sum_{\delta | Q_2} \sum_{\sigma < T\delta^{-1}} \delta^{\frac{1}{2}} \leqslant \sum_{\delta | Q_2} T \delta^{-\frac{1}{2}} \ll T Q_2^{\varepsilon},$$

this expression is

$$\ll Q_2^{\varepsilon_1} (1 + q^{\frac{m_2}{m}} d^{-m_2} Q_2^{-m_2}) d^{-n} q^{\frac{n}{m}} . \tag{32}$$

Now we are able to complete the estimation of the summation (17). By (21) and (28) we have

$$|\Gamma_{q_1}(S_1)| \leqslant 2^{n\alpha(q_1)} Q^{n-1} Q_2^{\frac{1}{2}},$$

with $q_1 = qd^{-1}$, $Q = q_1 2^{-\alpha(q_1)}$, $2^{\alpha(q_1)} \| q_1$, $Q_2 | Q$. We apply the inequality (32) for the number of possible quadratic forms S_1 and sum it over all values of Q_2. Thus we have

$$\sum_q \ll \sum_{d|q} d^n \sum_{Q_2 | qd^{-1}} 2^{n\alpha(qd^{-1})} Q^{n-1} Q_2^{\frac{1}{2}+\varepsilon_1}$$

$$\times d^{-n} q^{\frac{n}{m}} (1 + q^{\frac{m_2}{m}} d^{-m_2} Q_2^{-m_2})$$

with $Q = qd^{-1}2^{-\alpha(qd^{-1})}$, $d|q$. Consequently, $2^{n\alpha(qd^{-1})}Q^{n-1} = 2^{\alpha(qd^{-1})}\times (qd^{-1})^{n-1}$, and hence we obtain

$$\sum\nolimits_q \ll 2^{\alpha(q)} q^{n-1+\frac{n}{m}+\varepsilon_2} \sum_{Q_2|q} Q_2^{\frac{1}{2}}\left(1+Q_2^{-m_2} q^{\frac{m_2}{m}}\right)$$

$$\ll 2^{\alpha(q)} q^{n-1+\frac{n}{m}+\varepsilon_2}\left(q^{\frac{1}{2}+\varepsilon}+q^{\frac{m_2}{m}+\varepsilon}\right).$$

$$\ll 2^{\alpha(q)}\left(q^{n-\frac{1}{2}+\frac{n}{m}+\varepsilon_3}+q^{n-1+\frac{n}{m}+\frac{m_2}{m}+\varepsilon_3}\right). \tag{33}$$

Since

$$\frac{n}{m}=\frac{2}{n+3}\leqslant\frac{1}{5},\quad \frac{n}{m}+\frac{m_2}{m}=\frac{n+1}{n+3}<1 \quad (n\geqslant 2),$$

it follows that

$$\sum\nolimits_q \ll 2^{\alpha(q)} q^n, \tag{34}$$

and therefore, by (14),

$$N(q) \ll q^{n-\frac{m_1}{m}} + 2^{\alpha(q)} q^{n-\frac{m_1}{m}} = (2^{\alpha(q)}+1)\, q^{n+\frac{n}{m}-1},$$

which implies the inequality (10) and thus completes the proof of the theorem.

Actually, the inequality (33) is slightly stronger than (34), since the former implies

$$\sum\nolimits_q \ll 2^{\alpha(q)} q^{n+\varepsilon_3-\min\left(\frac{1}{2}-\frac{n}{m},\ 1-\frac{n+m_2}{m}\right)}$$

$$= 2^{\alpha(q)} q^{n+\varepsilon_3-\delta_0}, \quad \delta_0=\frac{n}{m}>0.$$

Consequently, we have

$$N(q) \ll q^{n-\frac{m_1}{m}} + 2^{\alpha(q)} q^{n-\frac{m_1}{m}-\delta_0} \tag{35}$$

This remark justifies the following sharpened version of the theorem:

Let $\psi(q)$ be a positive function for which the series

$$\sum_{q=1}^{\infty} \psi^n(q)\, q^{\frac{n}{m}-1} \tag{36}$$

converges and has a monotonically decreasing sequence of elements. Then the system of inequalities

$$\max(\|\omega_i q\|, \|\omega_i\omega_j q\|) < \psi(q) \quad (i, j = 1, 2, \ldots, n)$$

has, for almost all $\bar{\omega}$, at most finitely many solutions.

(By the principle of transfer an analogous improvement can be obtained for the inequality

$$\|\Phi(\omega_1, \ldots, \omega_n)\| < \Psi(h)$$

with a suitable function $\Psi(h)$ which decreases more slowly than $h^{-m-\delta}$, $\delta > 0$.)

Since the sequence of elements of the series (36) is monotonically decreasing, we have

$$\psi(q) < \tau q^{-\frac{1}{m}}, \quad q > q_0(\tau), \tag{37}$$

and again we can apply the preceding argument in order to obtain the inequality (35). Consequently, the problem reduces itself to the question of whether the series

$$\sum_{q=1}^{\infty} \frac{N(q)}{q^n} \psi^n(q)$$

is convergent. This is indeed the case, since by (35), (36) and (37) we have

$$\sum_{q=1}^{\infty} \frac{N(q)}{q^n} \psi^n(q) \ll \sum_{q=1}^{\infty} \left(q^{n-\frac{m_1}{m}} + 2^{\alpha(q)} q^{n-\frac{m_1}{m}-\delta_0}\right) \frac{\psi^n(q)}{q^n}$$

$$= \sum_{q=1}^{\infty} \psi^n(q) q^{-\frac{m_1}{m}} + \sum_{q=1}^{\infty} 2^{\alpha(q)} q^{-\frac{m_1}{m}-\delta_0} \psi^n(q)$$

$$\ll \sum_{q=1}^{\infty} \psi^n(q) q^{\frac{n}{m}-1} + \sum_{q=1}^{\infty} 2^{\alpha(q)} q^{-\frac{m_1}{m}-\delta_0} q^{-\frac{n}{m}}$$

$$\ll 1 + \sum_{q=1}^{\infty} 2^{\alpha(q)} q^{-1-\delta_0} \ll 1.$$

By the same method we can deal with the inequality

$$\prod_{1 \leq i < j \leq n} \max(\|\omega_i q\|, \|\omega_j q\|, \|\omega_i\omega_j q\|) < q^{-\frac{n+1}{n+3}-\delta} \tag{38}$$

In fact, we may assert that this inequality has, for almost all $\bar{\omega}$, at most finitely many solutions if we can show that the system of inequalities

$$\max_{(i,j)} (\| \omega_i q \|, \| \omega_j q \|, \| \omega_i \omega_j q \|) < q^{-r_{ij}-\delta} \tag{39}$$

has, for almost all $\bar{\omega}$, at most finitely many solutions provided that the numbers r_{ij} satisfy the conditions

$$r_{ij} = r_{ji} \geqslant 0, \quad \sum_{1 \leqslant i \leqslant j \leqslant n} r_{ij} = \frac{n+1}{n+3} \tag{40}$$

(In this case we let

$$\max (\| \omega_i q \|, \| \omega_j q \|, \| \omega_i \omega_j q \|) = q^{-\rho_{ij}},$$

and we introduce approximating ϵ-nets for the quantities ρ_{ij}.) By (39) we have

$$\max \left(\left| \omega_i - \frac{a_i}{q} \right|, \; \left| \omega_i - \frac{a_j}{q} \right|, \; \left| \omega_i \omega_j - \frac{a_{ij}}{q} \right| \right) < q^{-1-r_{ij}-\delta}$$

with suitable integers a_i, a_j, and a_{ij}. Therefore,

$$\left\| \frac{a_i a_j}{q} \right\| < 4\, q^{-r_{ij}-\delta} \quad (1 \leqslant i \leqslant j \leqslant n), \tag{41}$$

and we have to find a lower bound for the number $N(q)$ of solutions of this system by integers a_i, $0 \leq a_i \leq q$ $(i = 1, 2, \cdots, n)$. Then we apply the quadratic forms (12), choosing the coefficients c_{ij} such that

$$|c_{ij}| \ll q^{r_{ij}} \quad (i, j = 1, 2, \ldots, n). \tag{42}$$

Thus we obtain the estimate

$$N(q) \ll q^{-\frac{n+1}{n+3}} \sum_S |\Gamma_q(S)|,$$

where the summation is taken over all quadratic forms S which satisfy (42). Consequently, we have

$$N(q) \ll q^{n - \frac{n+1}{n+3}} + q^{-\frac{n+1}{n+3}} \sum_{S \neq (0)} |\Gamma_q(S)|. \tag{43}$$

The last summation may be estimated by the same reasoning which we have used before. The only modification needed concerns the estimation of the numbers c'_{ij} which satisfy the condition (29). If the numbers c'_{ij} $(i = 1, 2, \cdots, n)$ are fixed then the number of possible choices of the vector $\bar{\alpha} = (\alpha_1, \alpha_2, \cdots, \alpha_n) \bmod Q_2$ is bounded by the bound given in (31), and for fixed α_i, α_j the number of possible values of c'_{ij} is $\ll 1 + q^{r_{ij}} d^{-1} Q_2^{-1}$. Therefore, the total number of different values of c'_{ij} $(1 \leq i < j \leq n)$ for a given $\bar{\alpha}$ is

$$\ll \prod_{1 \leq i < j \leq n} (1 + q^{r_{ij}} d^{-1} Q_2^{-1}).$$

Consequently, the number of possible matrices c'_{ij} $(i, j = 1, 2, \cdots, n)$ is bounded by the expression

$$\sum_{|c_{11}| \ll q^{r_{11}} d^{-1}} \cdots \sum_{|c_{nn}| \ll q^{r_{nn}} d^{-1}} \prod_{i=1}^{n} Q_2^{\epsilon} (c'_{ii}, Q_2)^{\frac{1}{2}}$$

$$\times \prod_{1 \leq i < j \leq n} (1 + q^{r_{ij}} d^{-1} Q_2^{-1})$$

$$\ll Q_2^{\varepsilon_1} d^{-n} q^{r_1} (1 + q^{r_2} Q_2^{-1}),$$

where we have introduced the abbreviations

$$r_1 = \sum_{i=1}^{n} r_{ii}, \quad r_2 = \sum_{1 \leq i < j \leq n} r_{ij}.$$

Now an application of (21) and (28) yields

$$\sum_{S \neq (0)} |\Gamma_q(S)|$$

$$\ll \sum_{d|q} d^n \sum_{Q_2 | qd^{-1}} 2^{n\alpha(qd^{-1})} Q^{n-1} Q_2^{\frac{1}{2}+\varepsilon_1} d^{-n} q^{r_1} (1 + q^{r_2} Q_2^{-1})$$

$$\ll 2^{\alpha(q)} q^{n-1+r_1+\varepsilon_2} \sum_{Q_2 | q} Q_2^{\frac{1}{2}} (1 + q^{r_2} Q_2^{-1})$$

$$\ll 2^{\alpha(q)} q^{n-1+r_1+\varepsilon_2} (q^{\frac{1}{2}+\varepsilon} + q^{r_2+\varepsilon}).$$

Consequently, it follows from (43) that

$$N(q) \ll q^{n - \frac{n+1}{n+3}} + 2^{\alpha(q)} q^{n-1-\frac{n+1}{n+3}+r_1+\varepsilon_2} (q^{\frac{1}{2}+\varepsilon} + q^{r_2+\varepsilon}) \ll$$

$$\ll 2^{\alpha(q)} q^{n-\frac{1}{2}-\frac{n+1}{n+3}+r_1+\varepsilon_3} + q^{n-\frac{n+1}{n+3}},$$

since $r_1 + r_2 = (n+1)/(n+3) < 1$. Thus, letting $r_3 = \sum_{i=1}^n \max_{j=1,\cdots,n} r_{ij}$, the measure of the set of possible points $\bar{\omega}$ is, for a fixed q,

$$\ll N(q) \prod_{i=1}^n q^{-1-\max_{(j)} r_{ij}-\delta} = N(q) q^{-n-r_3-n\delta},$$

$$\ll 2^{\alpha(q)} q^{-\frac{n+1}{n+3}-r_3-\delta} + 2^{\alpha(q)} q^{-\frac{1}{2}-\frac{n+1}{n+3}-r_3+r_1\delta} \leqslant 2^{\alpha(q)} q^{-1-\delta},$$

assuming the inequalities

$$\frac{n+1}{n+3} + r_3 \geqslant 1, \quad \frac{n+1}{n+3} + r_3 \geqslant \frac{1}{2} + r_1.$$

to be true. The second one of these inequalities is trivial since $(n+1)/(n+3) > ½$ $(n = 2, 3, \cdots)$ and since

$$r_3 = \sum_{i=1}^n \max_{(j)} r_{ij} \geqslant \sum_{i=1}^n r_{ii} = r_1,$$

due to the fact that $\max_{j=1,\cdots,n} r_{ij} \geq r_{ii}$ $(i = 1, \cdots, n)$. The first inequality is equivalent to $r_3 \geq 2/(n+3)$. To show that this inequality is in fact true, we observe that $r_2 = (n+1)/(n+3) - r_1$, and hence

$$\sum_{1 \leqslant i,j \leqslant n} r_{ij} = r_1 + 2r_2 = \frac{2(n+1)}{n+3} - r_1.$$

On the other hand,

$$\max_{(j)} r_{ij} \geqslant \frac{1}{n} \sum_{j=1}^n r_{ij},$$

and therefore

$$r_3 = \sum_{i=1}^n \max_{j=1,\cdots,n} r_{ij} \geqslant \frac{1}{n} \sum_{1 \leqslant i,j \leqslant n} r_{ij} = \frac{2}{n+3}$$

$$+ \frac{1}{n}\left(\frac{2}{n+3} - r_1\right) \geqslant \frac{2}{n+3},$$

provided that $r_1 \leq 2/(n+3)$. However, if $r_1 > 2/(n+3)$, then trivially $r_3 \geq r_1 > 2/(n+3)$.

Thus, again using the convergence of the series $\sum_{q=1}^{\infty} 2^{\alpha(q)} q^{-1-\delta}$, we may conclude that, for almost all $\bar{\omega}$, the inequality (38) has, given any $\delta > 0$, at most finitely many integral solutions q. Conversely, it follows from Minkowski's lemma on linear forms that, for any $\bar{\omega}$, the left-hand side of (38) is infinitely often less than or equal to

$$\prod_{1 \leq i < j \leq n} q^{-\frac{1}{m}} = q^{-\frac{m_1}{m}} = q^{-\frac{n+1}{n+3}}$$

Consequently, the inequality (38) with $\delta = 0$ has infinitely many solutions.

BIBLIOGRAPHY

[1] N. C. Ankeny and S. Chowla, *A note on the class-number of real quadratic fields*, Acta Arith. 6 (1960), 145–147. MR 22 #6780.

[2] V. I. Arnol'd, *Small denominators and problems of stability of motion in classical and celestial mechanics*, Uspehi Mat. Nauk 18 (1963), no. 6 (114), 91–192 = Russian Math. Surveys 18 (1963), no. 6, 85–191. MR 30 #943.

[3] Z. I. Borevič and I. R. Šafarevič, *Number theory*, "Nauka", Moscow, 1964; English transl., Pure and Appl. Math., vol. 20, Academic Press, New York, 1966. MR 30 #1080; MR 33 #4001.

[4] J. W. S. Cassels, *Some metrical theorems in Diophantine approximation*. I, Proc. Cambridge Philos. Soc. 46 (1950), 209–218. MR 12, 162.

[5] ———, *Some metrical theorems in Diophantine approximation*. V: *On a conjecture of Mahler*, Proc. Cambridge Philos. Soc. 47 (1951), 18–21. MR 12, 679.

[6] ———, *An introduction to Diophantine approximation*, Cambridge Tracts in Math. and Math. Phys., no. 45, Cambridge Univ. Press, New York, 1957; Russian transl., IL, Moscow, 1961. MR 19, 396.

[7] N. G. Čudakov, *Introduction to the theory of Dirichlet's L-functions*, OGIZ, Moscow, 1947. (Russian) MR 11, 234.

[8] H. Davenport, *On the class-number of binary cubic forms*. I, II, J. London Math. Soc. 26 (1951), 183–192; 192–198. MR 13, 323.

[9] ———, *A note on binary cubic forms*, Mathematika 8 (1961), 58–62. MR 24 #A1253.

[10] B. N. Delone and D. K. Faddeev, *Theory of irrationalities of the third degree*, Trudy Mat. Inst. Steklov. 11 (1940); English transl., Transl. Math. Monographs, vol. 10, Amer. Math. Soc., Providence, R. I., 1964. MR 2, 349.

[11] R. J. Duffin and A. C. Schaeffer, *Khintchine's problem in metric Diophantine approximation*, Duke Math. J. 8 (1941), 243–255. MR 3, 71.

[12] N. I. Fel'dman, *The approximation of certain transcendental numbers*, Dokl. Akad. Nauk SSSR 66 (1949), 565–567. (Russian) MR 11, 232.

[13] ———, *Approximation of certain transcendental numbers.* I: *The approximation of logarithms of algebraic numbers*, Izv. Akad. Nauk SSSR Ser. Mat. 15 (1951), 53–74; English transl., Amer. Math. Soc. Transl. (2) **59** (1966), 224–245. MR 12, 595.

[14] A. O. Gel'fond, *Transcendental and algebraic numbers*, GITTL, Moscow, 1952; English transl., Dover, New York, 1960. MR 15, 292; MR 22 #2598.

[15] A. O. Gel'fond and Ju. V. Linnik, *On Thue's method in the problem of effectiveness in quadratic fields*, Dokl. Akad. Nauk SSSR 61 (1948), 773–776. (Russian) MR **10**, 354.

[16] P. R. Halmos, *Measure theory*, Van Nostrand, Princeton, N. J., 1950; Russian transl., IL, Moscow, 1953. MR 11, 504.

[17] H. Heilbronn and E. H. Linfoot, *On the imaginary quadratic corpora of class-number one*, Quart. J. Math. Oxford Ser. 5 (1934), 293–301.

[18] A. Ja. Hinčin [A. Khintchine], *Zwei Bermerkungen zu einer Arbeit des Herrn Perron*, Math. Z. 22 (1925), 274–284.

[19] ———, *Zur metrischen Theorie der Diophantischen Approximationen*, Math. Z. **24** (1926), 706–714.

[20] ———, *Über eine Klasse linearer Diophantischer Approximationen*, Rend. Circ. Mat. Palermo **50** (1926), 170–195.

[21] ———, *Continued fractions*, 3rd ed., Fizmatgiz, Moscow, 1961; English transl., Univ. of Chicago Press, Chicago, Ill., 1964. MR **28** #5037.

[22] Hau Loo-keng, *Die Abschätzung von Exponentialsummen und ihre Anwendung in der Zahlentheorie*, Teubner, Leipzig, 1959; Russian transl., "Mir", Moscow, 1964. MR 24 #A94; MR 31 #4768.

[23] F. Kasch, *Über eine metrische Eigenschaft der S-Zahlen*, Math. Z. 70 (1958), 263–270. MR 21 #1296.

[24] ———, *Ein metrischer Beitrag über Mahlersche S-Zahlen.* II, J. Reine Angew. Math. 203 (1960), 157–159. MR 23 #A2380.

[25] F. Kasch and B. Volkmann, *Zur Mahlerschen Vermutung über S-Zahlen*, Math. Ann. 136 (1958), 442–453. MR 21 #1297.

[26] ———, *Metrische Sätze über transzendente Zahlen in p-adischen Körpern*, Math. Z. 72 (1959/60), 367–378. MR **23** #A3713.

[27] ———, *Metrische Sätze über transzendente Zahlen in p-adischen Körpern. II*, Math. Z. 78 (1962), 171–174. MR 25 #2035.

[28] K. Knopp, *Mengentheoretische Behandlung einiger Probleme der Diophantischen Approximationen und der transfiniten Wahrscheinlichkeiten*, Math. Ann. 95 (1926), 409–429.

[29] J. F. Koksma, *Diophantische Approximationen*, Ergebnisse der Math. und ihrer Grenzgebiete, vol. 4, Springer-Verlag, Berlin, 1936.

[30] ———, *Über die Mahlersche Klasseneinteilung der transzendenten Zahlen und die Approximation komplexer Zahlen durch algebraische Zahlen*, Monatsh. Math. Phys. 48 (1939), 176–189. MR 1, 137.

[31] J. P. Kubilius, *On an application of I. M. Vinogradov's method to the solution of a problem of the metric theory of numbers*, Dokl. Akad. Nauk SSSR 67 (1949), 783–786. (Russian) MR 11, 82.

[32] ———, *On a metrical problem in the theory of Diophantine approximation*, Trudy Akad. Nauk Litov. SSR Ser. B, 1959, no. 2 (18), 3–7. (Russian) R Ž Mat. 1960 #6124.

[33] A. G. Kuroš, *Lectures in general algebra*, Fizmatgiz, Moscow, 1962; English transl., Internat. Series of Monographs in Pure and Appl. Math., vol. 70, Pergamon Press, New York, 1965. MR 25 #5097; MR 31 #3483.

[34] S. Lang, *Diophantine geometry*, Interscience Tracts in Pure and Appl. Math., no. 11, Interscience, New York, 1962. MR 26 #119.

[35] D. N. Lenskoĭ, *Functions in non-Archimedean normed fields*, Izdat. Saratov. Univ., Saratov, 1962. (Russian) MR 27 #2498.

[36] W. J. LeVeque, *Note on S-numbers*, Proc. Amer. Math. Soc. 4 (1953), 189–190. MR 14, 956.

[37] ———, *On Mahler's U-numbers*, J. London Math. Soc. 28 (1953), 220–229. MR 14, 956.

[38] D. J. Lock, *Metrisch-Diophantische Onderzoekingen in $K(P)$ en $K^{(n)}(P)$*, Dissertation, Vrije Univ., Amsterdam, 1947. MR 9, 79.

[39] M. Loève, *Probability theory*, 2nd ed., Van Nostrand, Princeton, N. J., 1960; Russian transl., IL, Moscow, 1962. MR 23 #A670.

[40] E. Lutz, *Sur les approximations diophantiennes linéaires p-adiques*, Actualités Sci. Indust., no. 1224, Hermann, Paris, 1955. MR 16, 1003.

[41] K. Mahler, *Zur Approximation der Exponentialfunction und des Logarithmus. I, II*, J. Reine Angew. Math. 166 (1932), 118–136; 137–150.

[42] ———, *Über das Mass der Menge aller S-Zahlen*, Math. Ann. 106 (1932), 131–139.

[43] ———, *Über diophantische Approximationen im Gebiete der P-adischen Zahlen*, Jber. Deutsch. Math. Verein. 44 (1934), 250–255.

[44] ———, *Über eine Klasseneinteilung der p-adischen Zahlen*, Mathematika 3 (1934/35), 177–185.

[45] ———, *Neuer Beweis eines Sätzes von A. Khintchine*, Mat. Sb. 1 (1936), 961–962.

[46] ———, *An analogue to Minkowski's geometry of numbers in a field of series*, Ann. of Math. (2) 42 (1941), 488–522. MR 2, 350.

[47] H. Minkowski, *Diophantische Approximationen*, Teubner, Leipzig, 1907.

[48] L. J. Mordell, *The Diophantine equation* $y^2 - k = x^3$, Proc. London Math. Soc. 13 (1914), 60–80.

[49] A. Ostrowski, *Über einige Lösungen der Funktionalgleichung* $f(x)f(y) = f(xy)$, Acta Math. 41 (1918), 271–284.

[50] A. G. Postnikov, *Arithmetic modeling of random processes*, Trudy Mat. Inst. Steklov. 57 (1960). (Russian) MR 26 #6146.

[51] W. Schmidt, *A metrical theorem in Diophantine approximation*, Canad. J. Math. 12 (1960), 619–631. MR 22 #9482.

[52] ———, *Bounds for certain sums; a remark on a conjecture of Mahler*, Trans. Amer. Math. Soc. 101 (1961), 200–210. MR 24 #A1883.

[53] ———, *Metrische Sätze über simultane Approximation abhängiger Grössen*, Monatsh. Math. 68 (1964), 154–166. MR 30 #1980.

[54] Th. Schneider, *Einführung in die transzendenten Zahlen*, Springer-Verlag, Berlin, 1957. MR 19, 252.

[55] V. G. Sprindžuk, *Metrical theorems on Diophantine approximation by algebraic numbers of bounded degree*, Dissertation, Leningrad. Gos. Univ., Leningrad, 1963. (Russian)

[56] ———, *On some general problems of approximating numbers by algebraic numbers*, Litovsk. Mat. Sb. 2 (1962), no. 1, 129–145. (Russian) MR 28 #1165.

[57] ———, *On theorems of Hinčin and Kubilius*, Litovsk. Mat. Sb. **2** (1962), no. 1, 147–152. (Russian) MR 28 #2088.

[58] ———, *Problems on the approximation of p-adic numbers*, Litovsk. Mat. Sb. 2 (1962), no. 1, 234. (Russian)

[59] ———, *On a conjecture of K. Mahler concerning the measure of the set of S-numbers*, Litovsk. Mat. Sb. 2 (1962), no. 2, 221–226. (Russian) MR 30 #1979.

[60] ———, *Metric theorems on algebraic approximation in the field of power series*, Litovsk. Mat. Sb. 2 (1962), no. 2, 207–213. (Russian) MR 30 #1978.

[61] ———, *On a classification of transcendental numbers*, Litovsk. Mat. Sb. **2** (1963), no. 2, 215–219. (Russian) MR 27 #5730.

[62] ———, *On algebraic approximations in the field of power series*, Vestnik Leningrad. Univ. Ser. Mat. Meh. Astronom. **18** (1963), no. 3, 130–134. (Russian) MR **28** #2089.

[63] ———, *On the number of solutions of the Diophantine equation* $x^3 = y^2 + A$, Dokl. Akad. Nauk BSSR 7 (1963), 9–11. (Russian) MR 27 #1413.

[64] ———, *On the measure of the set of S-numbers in the p-adic field*, Dokl. Akad. Nauk SSSR 151 (1963), 1292 = Soviet Math. Dokl. 4 (1963), 1201–1202. MR 29 #5784.

[65] ———, *On Mahler's conjecture*, Dokl. Akad. Nauk SSSR 154 (1964), 783–786 = Soviet Math. Dokl. 5 (1964), 183–187. MR **28** #2090a.

[66] ———, *More on Mahler's conjecture*, Dokl. Akad. Nauk SSSR 155 (1964), 54–56 = Soviet Math. Dokl. 5 (1964), 361–363. MR 28 #2090b.

[67] ———, *Proof of Mahler's conjecture on the measure of the set of complex S-numbers*, Uspehi Mat. Nauk **19** (1964), 191–194. (Russian)

[68] ———, *A proof of Mahler's conjecture on the measure of the set of S-numbers*, Izv. Akad. Nauk SSSR Ser. Mat. 29 (1965), 379–436; English transl., Amer. Math. Soc. Transl. (2) 51 (1966), 215–272. MR 31 #4762.

[69] H. Turkstra, *Metrische bijdragen tot de theorie der Diophantische approximaties in het lichaam der p-adische getallen*, Dissertation, Vrije Univ., Amsterdam, 1936.

[70] I. M. Vinogradov, *Elements of number theory*, 5th ed., GITTL, Moscow, 1949; English transl., Dover, New York, 1954. MR 12, 10; MR 15, 933.

[71] ———, *The method of trigonometrical sums in the theory of numbers*, Trudy Mat. Inst. Steklov. 23 (1947); English transl., Interscience, New York, 1954. MR **10**, 599; MR 15, 941.

[72] B. Volkmann, *Zum kubischen Fall der Mahlerschen Vermutung*, Math. Ann. 139 (1959), 87–90. MR **23** #A3711.

[73] ———, *Zur Mahlerschen Vermutung im Komplexen*, Math. Ann. 140 (1960), 351–359. MR 23 #A3712.

[74] ———, *Ein metrischer Beitrag über Mahlersche S-Zahlen.* I, J. Reine Angew. Math. 203 (1960), 154–156. MR 23 #A2379.

[75] ———, *The real cubic case of Mahler's conjecture*, Mathematika **8** (1961), 55–57. MR 24 #A1254.

[76] ———, *Zur metrischen Theorie der S-Zahlen*, J. Reine Angew. Math. 209 (1962), 201–210. MR 25 #5036.

[77] ———, *Zur metrischen Theorie der S-Zahlen.* II, J. Reine Angew. Math. 213 (1963), 58–65. MR 28 #61.

[78][1] B. L. van der Waerden, *Moderne algebra.* Vol. 1, 2nd rev. ed., Springer-Verlag, Berlin, 1937; reprint, Stechert, New York, 1943; Russian transl., GITTL, Moscow, 1947; English transl., Ungar, New York, 1949. MR **10**, 587.

[79] H. Weyl, *Algebraic theory of numbers*, Ann. of Math. Studies, no. 1, Princeton Univ. Press, Princeton, N.J., 1940; Russion transl., IL, Moscow, 1947. MR **2**, 37.

[80] E. Wirsing, *Approximation mit algebraischen Zahlen beschränkten Grades*, J. Reine Angew. Math. **206** (1960), 67–77. MR 26 #79.

[81] O. Zariski and P. Samuel, *Commutative algebra.* Vol. 1, Van Nostrand. Princeton, N. J., 1958; Russian transl., IL, Moscow, 1963. MR **19**, 833.

[82] Alan Baker, *On a theorem of Sprindžuk*, Proc. Roy. Soc. Ser. A 292 (1966), 92–104.

1) The references in the text are to these three editions. In some other editions the chapter and section numbering is quite different.

Date Due

Due	Returned	Due	Returned